AF389173

Advances in Control of Power Electronic Converters

Advances in Control of Power Electronic Converters

Editors

Oswaldo Lopez Santos
Germain García

MDPI • Basel • Beijing • Wuhan • Barcelona • Belgrade • Manchester • Tokyo • Cluj • Tianjin

Editors
Oswaldo Lopez Santos
Electronics Engineering
Universidad de Ibagué
Ibagué
Colombia

Germain García
Computer and Electrical
Engineering Department (GEI)
INSA
Toulouse
France

Editorial Office
MDPI
St. Alban-Anlage 66
4052 Basel, Switzerland

This is a reprint of articles from the Special Issue published online in the open access journal *Applied Sciences* (ISSN 2076-3417) (available at: www.mdpi.com/journal/applsci/special_issues/control_dc_ac_power_electronic_converters).

For citation purposes, cite each article independently as indicated on the article page online and as indicated below:

LastName, A.A.; LastName, B.B.; LastName, C.C. Article Title. *Journal Name* **Year**, *Volume Number*, Page Range.

ISBN 978-3-0365-1392-8 (Hbk)
ISBN 978-3-0365-1391-1 (PDF)

Contents

About the Editors

Oswaldo Lopez Santos

He graduated as an Electronic Engineer in 2006, obtained anMSc degree on Industrial Automation in 2010, and finished his PhD degree in Automation from the Institut National des Sciences Appliquées, Toulouse, France in 2015. He has 12 years of professional experience with outstanding performance as a professor and a researcher in the areas of control, power electronics and industrial electronics, also including five years as design engineer in the Colombian power electronics industry. Since 2009, he has been professor of the Universidad de Ibagué where He is currently Associate Professor of the Electronics Engineering Program. He is currently categorized as Associate Researcher by Minciencias and serves as a reviewer for several journals of recognized editorials. His research interest is focused on the control of power electronics converters for energy processing in applications related to autonomous systems and microgrids.

Germain García

He obtained his engineering degree in 1984, his PhD in automatic control in 1988 and the Habilitation to Conduct Research (HDR) in 1997 from the National Institute of Applied Sciences (INSA) in Toulouse—France, where he is currently a Professor of Exceptional Class dedicated to the area of control. Professor Garcia is a researcher in the Methods, Algorithms and Control (MAC) group of the Laboratory of Analysis and Architecture of Systems (LAAS) attached to the National Center for Scientific Research (CNRS) in France. His most representative results in the area of control include the direction of more than 20 PhD students, the authorship of two research books and one teaching book, and numerous papers published in journals and conferences, including papers in the journal Automatica and the IEEE Transactions. His research interests include robust and hybrid control theory, matrix inequalities, constrained control and perturbed singularity models.

Editorial

Special Issue "Advances in Control of Power Electronic Converters"

Oswaldo Lopez-Santos [1] and Germain Garcia [2,*]

[1] Facultad de Ingeniería, Universidad de Ibagué, Carrera 69 Calle 22 Barrio Ambalá, Ibagué 730001, Colombia; oswaldo.lopez@unibague.edu.co
[2] LAAS-CNRS, Université de Toulouse, CNRS, INSA, 31077 Toulouse, France
* Correspondence: garcia@laas.fr

Abstract: The use of power converters has grown in the last years with the advances in photovoltaic and wind based power generation systems, and the progress in modern concepts such as microgrids and electric mobility. A consequence has been the development of devices allowing for the exchange of energy among different distribution buses, and feeding AC or DC loads from low DC voltage levels, whose proper operation is achieved by means of specialized control systems. Simultaneously, the power converters used for conventional industrial applications have evolved thanks to the application of new control methods, and the combination of these with well-established techniques. This special issue contributes theoretical and practical advances to the state-of-the-art field at the crossroads of power electronics and control systems. The seven included papers cover particular applications requiring either DC–DC, DC–AC or AC–DC conversion stages.

Keywords: control of power converters; power electronics; control systems

Citation: Lopez-Santos, O.; Garcia, G. Special Issue "Advances in Control of Power Electronic Converters". *Appl. Sci.* **2021**, *11*, 4585. https://doi.org/10.3390/app11104585

Received: 2 May 2021
Accepted: 10 May 2021
Published: 18 May 2021

Publisher's Note: MDPI stays neutral with regard to jurisdictional claims in published maps and institutional affiliations.

1. Introduction

Despite conventional techniques for the control of power converters established for several decades, the emergence of new challenges and the search for improved performance to ensure better energy utilization continues to motivate continuous and growing research in this area. The published results over the last years, both theoretical and practical, show the important impact of control design techniques which, associated with the significant recent progresses seen in the domain of materials, electronic devices or components, offer new perspectives in a domain where the problems become more and more complex. The objective of this special issue is to stimulate research in the area of control of power electronic converters, and promote the emergence of methods justified by rigorous theoretical analysis and validated with the help of simulation tools and experimental development. Among the eleven submitted papers to this special issue, only seven papers have been retained. Among them, three papers are focused on the control of DC–DC converters, three on the control of DC–AC converters and one on the control of a three phase rectifier.

2. Control of DC-DC Converters

The paper of Torres-Pinzon et al. [1] uses Takagi–Sugeno (T–S) fuzzy controllers to improve disturbance rejection and to optimize the control effort, guaranteeing the large-signal stability of power converters in a broad operation domain. The paper develops T–S models of boost and buck-boost converters, which were selected because of their non-minimum phase type dynamic behavior. Design of the fuzzy controllers is performed by using the Parallel Distributed Compensation technique (PDC). The approach is validated by means of simulation results for both converters, and experimental results are presented for the boost converter using a laboratory prototype of 60 W. For this work, it is worth highlighting that proposed control is entirely implemented using analogue electronics, i.e., passive components, operational amplifiers and multipliers.

1

Nedia Aouani and Carlos Olalla propose in [2] a novel framework for application of robust linear quadratic regulator (LQR)-based control in DC–DC power converters. In the same vein of the previous work, the controller design is performed using Linear Matrix Inequalities (LMIs) and Lyapunov stability theory, leading to a control ensuring robust stability. In this context, the converter is described by a linear parameter-varying polytopic model, integrating both uncertainties and rate of change of the stated variables. The successful operation of this method has been validated by means of simulated results using a conventional boost converter, providing comparison with a controller obtained using the method previously published in [3]. The main contributions of the work are the possibility to enlarge the region of uncertain parameters in which stability is ensured, and the improved regulation performance and robustness.

The paper by Gonzalez-Castaño et al. analyzes the undesired impacts of quantization (limit cycle oscillation) in the coupled inductor buck-boost converter when a two-loop digital current control is used to ensure output voltage regulation [4]. The selected control architecture involves an inner loop of multi-sampled average current control and an outer loop of voltage regulation. This work integrates design constraints for control gains and signal quantization to avoid these effects when the outer voltage loop is added. A laboratory prototype of 400 V and 1.6 kW is used to illustrate the presence of the studied phenomenon and verify the correctness of the proposed design conditions when the control system is implemented into a DSP. The contribution of this paper is highly useful to dealing with practical issues of both design and implementation of digital controllers for DC–DC power converters.

3. Control of DC–AC Converters

The paper by El Aroudi et al. [5] studies the steady-behavior of a differential boost inverter used for generating a sinewave AC voltage from a DC source. The dynamics of the converter are analyzed using an accurate discrete time approach, adopting a quasi-static approximation and the Floquet theory. The undesired sub-harmonic oscillation exhibited by the inverter in some intervals is accurately predicted by means of the complement between analytical expressions and computational procedures. The study provides stability boundaries in terms of the proportional gain of the PI controller used to track the output voltage reference. The results contribute to the design of a boost inverter avoiding the consequences of the sub-harmonic oscillation in the quality of the input current and the output voltage. The proposed simulation results validate the theoretical predictions and the accuracy of the analysis.

This contribution to the special issue [6] is focused on the same boost inverter as before. The objective is to reduce the voltage stress in power semiconductors by enforcing voltage references in the capacitor of the converters with a predefined harmonic content, which helps to decrease the voltage level required in each leg of the inverter for the entire cycle of the output voltage. Signal analysis allows definition that the voltage references with two harmonics is enough to achieve a considerable reduction of the voltage on the power semiconductors. To guarantee a proper tracking of the desired reference, two separate multiloop controllers were implemented, both using inner current control and outer voltage control. A complete linear modelling of the converter using the proposed control approach was derived from application of the equivalent control method. Moreover, qualitative analysis of the converter variables employing the harmonic balance method allowed them to derive a simplified plant model to facilitate voltage control design. The paper also compares the use of the proposed controller with the one developed in [7], showing how a small increase in complexity allows enforcement that minimum value of the capacitor voltages is close to the input voltage, then further reducing the voltage on semiconductors. A complete set of simulation results are provided for two electric standards (220 V/50 Hz and 120 V/60 Hz).

Iqbal et al. propose in their paper [8] a novel dead time compensation method for improving power quality and efficiency of inverters feeding induction motors. The method is developed considering a three phase IGBT bridge feeding the load through an output inductor. An idealized nonlinear model of the inverter is used to obtain theoretical expressions defining the effect of the dead time on the variables of the inverter. The converter is controlled using a constant frequency Pulse Width Modulator (PWM). The validity of the approach is confirmed by means of simulation results using a common V/f strategy for speed variation of two case studies. The results show that the proposed method reduces Total Harmonic Distortion (THD), increasing the quality of the output current.

4. Control of AC–DC Converters

The paper presented in the field of AC–DC converters by Ortiz-Castrillon et al. puts forward a single surface sliding mode control, with an adaptive hysteresis band developed for the semi-bridgeless boost type rectifier [9]. The proposed control uses a single sliding surface which considerably differs to the classic cascade control architecture (a PI controller regulating the output voltage in the outer loop and a current controller to track appropriate references in the inner loop). The proposal integrates into the sliding surface a normalized term for the output voltage, one term for the current error and one term for its integral. All the conditions required to ensure stability of the sliding motion are validated, supporting the proposal theoretically, including start-up and large perturbation conditions. The paper provides both simulation and experimental results, validating the correct operation of the controller regarding tracking of the current reference, voltage regulation and disturbance rejection.

Author Contributions: Both authors have contributed equally to the development of this editorial. Both authors have read and agreed to the published version of the manuscript.

Funding: This editorial received no external funding.

Institutional Review Board Statement: Not applicable.

Informed Consent Statement: Not applicable.

Data Availability Statement: Not applicable.

Acknowledgments: The invited editors want to express their gratitude to all authors contributing their work to this special issue. Also, they extend their sincere appreciation of the hard work of reviewers and editors of the editorial team of Applied Sciences.

Conflicts of Interest: The author declares no conflict of interest.

References

1. Torres-Pinzón, C.A.; Paredes-Madrid, L.; Flores-Bahamonde, F.; Ramirez-Murillo, H. LMI-Fuzzy Control Design for Non-Minimum-Phase DC-DC Converters: An Application for Output Regulation. *Appl. Sci.* **2021**, *11*, 2286. [CrossRef]
2. Aouani, N.; Olalla, C. Robust LQR Control for PWM Converters with Parameter-Dependent Lyapunov Functions. *Appl. Sci.* **2020**, *10*, 7534. [CrossRef]
3. Olalla, C.; Leyva, R.; el Aroudi, A.; Queinnec, I. Robust LQR Control for PWM Converters: An LMI Approach. *IEEE Trans. Ind. Electron.* **2009**, *56*, 2548–2558. [CrossRef]
4. González-Castaño, C.; Restrepo, C.; Giral, R.; Vidal-Idiarte, E.; Calvente, J. ADC Quantization Effects in Two-Loop Digital Current Controlled DC-DC Power Converters: Analysis and Design Guidelines. *Appl. Sci.* **2020**, *10*, 7179. [CrossRef]
5. El Aroudi, A.; Al-Numay, M.; Haroun, R.; Huang, M. Analysis of Subharmonic Oscillation and Slope Compensation for a Differential Boost Inverter. *Appl. Sci.* **2020**, *10*, 5626. [CrossRef]
6. López-Santos, O.; García, G. Double Sliding-Surface Multiloop Control Reducing Semiconductor Voltage Stress on the Boost Inverter. *Appl. Sci.* **2020**, *10*, 4912. [CrossRef]
7. Flores-Bahamonde, F.; Valderrama-Blavi, H.; Bosque-Moncusi, J.M.; García, G.; Martínez-Salamero, L. Using the sliding-mode control approach for analysis and design of the boost inverter. *IET Power Electron.* **2016**, *9*, 1625–1634. [CrossRef]
8. Iqbal, S.; Xin, A.; Jan, M.U.; Abdelbaky, M.A.; Rehman, H.U.; Salman, S.; Aurangzeb, M.; Rizvi, S.A.A.; Shah, N.A. Improvement of Power Converters Performance by an Efficient Use of Dead Time Compensation Technique. *Appl. Sci.* **2020**, *10*, 3121. [CrossRef]
9. Ortiz-Castrillón, J.R.; Mejía-Ruiz, G.E.; Muñoz-Galeano, N.; López-Lezama, J.M.; Cano-Quintero, J.B. A Sliding Surface for Controlling a Semi-Bridgeless Boost Converter with Power Factor Correction and Adaptive Hysteresis Band. *Appl. Sci.* **2021**, *11*, 1873. [CrossRef]

Article

LMI-Fuzzy Control Design for Non-Minimum-Phase DC-DC Converters: An Application for Output Regulation

Carlos Andrés Torres-Pinzón [1,*,†], Leonel Paredes-Madrid [2,*,†], Freddy Flores-Bahamonde [3,*,†] and Harrynson Ramirez-Murillo [4,†]

1 MEM (Modelling-Electronics and Monitoring Research Group), Faculty of Electronic Engineering, Universidad Santo Tomás, Cra 9 No. 51-11, Bogotá 110311, Colombia
2 Faculty of Mechanic, Electronic and Biomedical Engineering, Universidad Antonio Nariño, Tunja 150002, Colombia
3 Department of Engineering Science (DCI), Universidad Andres Bello, Av. Antonio Varas 880, Providencia, Santiago 7500971, Chile
4 Department of Electrical Engineering, Universidad de La Salle, Cra. 2 No. 10-70, Candelaria, Bogotá 111711, Colombia; haramirez@unisalle.edu.co
* Correspondence: carlostorresp@usantotomas.edu.co (C.A.T.-P.); paredes.leonel@uan.edu.co (L.P.-M.); freddy.flores@unab.cl (F.F.-B.)
† These authors contributed equally to this work.

Abstract: Robust control techniques for power converters are becoming more attractive because they can meet with most demanding control goals like uncertainties. In this sense, the Takagi-Sugeno (T-S) fuzzy controller based on linear matrix inequalities (LMI) is a linear control by intervals that has been relatively unexplored for the output-voltage regulation problem in switching converters. Through this technique it is possible to minimize the disturbance rejection level, satisfying constraints over the decay rate of state variables as well as the control effort. Therefore, it is possible to guarantee, a priori, the stability of the large-signal converters in a broad operation domain. This work presents the design of a fuzzy control synthesis based on a T-S fuzzy model for non-minimum phase dc-dc converters, such as boost and buck-boost. First, starting from the canonical bilinear converters expression, a Takagi-Sugeno (T-S) fuzzy model is obtained, allowing to define the fuzzy controller structure through the parallel distributed compensation technique (PDC). Finally, the fuzzy controller design based on LMIs is solved for the defined specification in close loop through MATLAB toolbox LMI. Simulations and experimental results of a 60 W prototype are presented to verify theoretical predictions.

Keywords: bilinear model; boost converter; Linear Matrix Inequalities (LMI); LMI-Fuzzy control; Takagi-Sugeno (T-S) fuzzy model

Citation: Torres-Pinzón, C.A.; Paredes-Madrid, L.; Flores-Bahamonde, F.; Ramirez-Murillo, H. LMI-Fuzzy Control Design for Non-Minimum-Phase DC-DC Converters: An Application for Output Regulation. *Appl. Sci.* **2021**, *11*, 2286. https://doi.org/10.3390/app11052286

Academic Editor: Ioannis Dassios

Received: 28 January 2021
Accepted: 25 February 2021
Published: 4 March 2021

Publisher's Note: MDPI stays neutral with regard to jurisdictional claims in published maps and institutional affiliations.

1. Introduction

DC-DC switching converters are one of the most classical power electronics circuits used to adapt non-regulated sources to different load requirements in many applications [1]. Hence, from telecommunication and computers to renewable energies equipment in which the voltages or currents controls are required, a DC-DC converter is necessary. In fact, during the last years dc-dc converters have been improved constantly, providing high performance at very high switching frequencies for a broad input and output voltages.

Nonetheless, based on the higher non-linear nature, or even due to the existence of parametric uncertainties, among others, switching converters present important dynamic complexities. Thus, to deal with these non-linearity issues and reach a good voltage regulation performance, classic linear feedback approaches have been commonly applied [1–3]. Though linear feedback control methods are accepted by the industry, it has been proven that these classic strategies also present malfunctioning or unstable behaviors under large disturbances. Basically, the latter is result from the fact that classical approaches do not consider the non-linearities of the converter. With the aim to overcome these limitations, researchers have been exploring other control techniques that contemplate the non-linearities and uncertainty parameters in switching DC-DC converters. For instance, robust control has been widely used during the last years attracting the interest in the power electronics field [4–6]. Unlike non-linear control and conventional linear control, robust control ensures a minimum of features regarding uncertainties in DC-DC converters.

In [7,8], the design of a robust state-feedback control for a minimum-phase buck converter is proposed. The works showed the use of simple conditions for LMIs guaranteeing the stability of the system as well as the disturbance rejection for the close-loop control. Nevertheless, they do not describe any experimental results and only verify the effectiveness of the proposed method through simulations. Similarly, in [9] it is also presented the design of a LMI-based robust control law by state-feedback, but in this case, applied to a non-minimum phase boost converter. This control method takes into consideration uncertainties and non-linearities of the converter, which are modeled like a convex polytope. This allows LMIs constraints to robustly guarantee a certain level of disturbances rejection and a specific location region for the poles. Contrarily to [7–9], presents precise experimental results in a boost prototype which are in agreement with the design requirements in spite of the uncertainties. Finally, in [10–12] other powerful LMI approaches are proposed for a boost regulator control, which allows to consider uncertainty in the converter and ensure its stability among different operation points due to the inclusion of a bilinear dynamic. The accuracy of these approaches is verified through an experimental prototype, which shows good similarity with theoretical predictions.

Moreover, over the last decades, the application of fuzzy logic in control systems has drawn the attention of the scientific community as well as numerous technicians of industrial processes. This methodology can be very useful when processes are rather complex for their analysis through conventional techniques, or when the available information is inaccurately or uncertainly interpreted. Despite the advantages of the fuzzy logic based controls there is a lack of tools for the prediction of the stability and robustness for closed-loop systems, generating some criticism in the automatic control field. For this reason, in the last few years, the design of more accurately fuzzy controllers, particularly the Takagi-Sugeno (T-S) based model, have registered significant breakthroughs thanks to the Linear Matrix Inequalities theory (LMI) and Lyapunov stability analysis [13,14].

In control engineering, the Model-Based Fuzzy Control (MBFC), which uses the concept of parallel distributed compensation, has been catalogued as an effective and systematic approach for highly non-linear control systems. Based on linear matrix inequalities and convex optimization techniques, this approach allows to secure better stability, performance and robustness properties and some established benefits over a wide operation region [14–16]. Thus, applied to DC-DC converters, the first references are found in [17,18]. In [17], a LMI-based integral fuzzy control for voltage regulation of a basic buck converter with zero-voltage switching (ZVS) is proposed, guaranteeing exponential stability in a broad field of operation. Afterwards in [18] a design methodology of a fuzzy control is presented for a boost converter, ensuring large-signal stability. In particular, through Lyapunov theory, the sufficient-stability conditions in terms of LMIs are adopted considering uncertainty in the parameters. In addition to this, relaxed conditions in the belonging functions are proposed with the purpose of relieving the conservatism of the presented approach. Nonetheless, this work does not describe any experimental prototype and only verifies the effectiveness of the method for a boost converter through simulations. Finally, in [19] an H_∞ fuzzy-control design method is presented for a DC-DC buck converter with input restrictions. Likewise, this work does not show experimental results.

Based on the previous statements, the motivation of this work consists in the use of a simple MBFC methodology for the output voltage regulation of two non-minimum phase DC-DC converters, such as boost and buck-boost. In [20,21] first designs of control synthesis proposed for DC-DC converters through simulations were performed in Matlab. Therefore, through simulations, it is proven that it is possible to ensure the constraints compliance such as: decay rate and effort control. Besides, starting from the obtained results, it is possible to observe at the same time that a minimum attenuation level is guaranteed between a load current disturbance and regulated output voltage. It is worth noticing, that through the application of the proposed strategy it is possible to consider the saturation of the duty cycle, which allows to attenuate the response of the control signal and maintain within the proper interval. Finally, through a simple and analogue implementation, experimental results of the boost converter are presented, verifying the advantages of the proposed method compared to a non-fuzzy LMI robust control.

The rest of the paper is organized as follows. In Section 2, MBFC theory is discussed, which will be used to build the T-S fuzzy models based on buck-boost and boost converters models. Then, in Section 3 the LMIs requirements for design are presented, taking into account the concept of CDP. Then, in Section 4, two examples of fuzzy control based on LMI design are presented for the output voltage regulation problem of buck-boost and boost converters, guaranteeing a certain level of disturbance rejection, a decay rate and control effort limitation. In addition, in this section it is presented a comparison of the proposed controller and a non-fuzzy LMI controller. Experimental results for the boost converter case are also shown in Section 4. Finally, in Section 5 the main conclusions are presented.

2. Takagi-Sugeno Fuzzy Modeling of DC-DC Converters

The construction of a mathematical model to describe the dynamic of a system in study is not easy. Since the model must include all the relevant characteristics associated with the dynamic. In addition, the mathematical expression is complex due to the non-linear nature of the system, making that a good approximation becomes essential. T-S fuzzy approach is a modeling methodology that considers the dynamics of a system as real as the exact model, through the implementation of several linear models. The main characteristic of this methodology is the formulation of the local dynamics of each fuzzy implication by a linear model [22]. The complete fuzzy model of the system is achieved by fuzzy blending of the linear system models. In this section, the averaged bilinear model and the Takagi-Sugeno fuzzy representation in a buck-boost and a boost converter are introduced.

2.1. T-S Fuzzy Model of a Buck-Boost Converter

Figure 1 shows the well-known circuit of a buck-boost converter, which is capable to step-up/down the output voltage $v_o(t)$ from the input one V_g. R represents the converter nominal load, while L and C stand for the inductance and capacitance values, respectively. Besides, source $i_o(t)$ represents the load current disturbance.

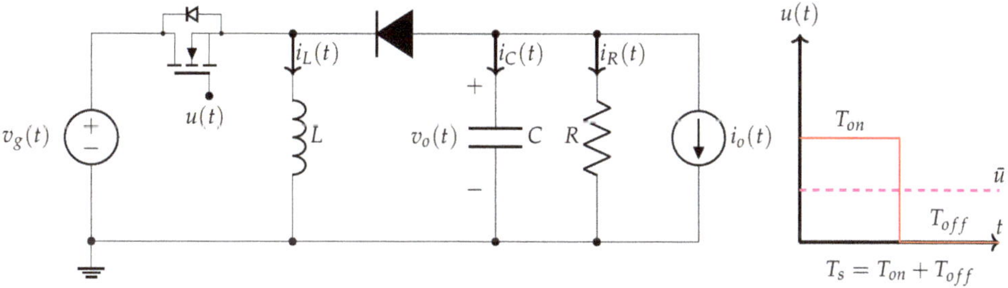

Figure 1. Schematic circuit of a buck-boost converter.

The state variables considered for the analysis are the inductor current $i_L(t)$ and the capacitor voltage $v_o(t)$. The binary signal $u(t)$ turn the MOSFET ON when $u = 1$ and OFF when $u = 0$ at a constant switching frequency $\frac{1}{T_s}$, such as shown in Figure 1. The step-up/down operation of the converter in steady state depends of the ratio $\frac{T_{on}}{T_s}$, which represents the duty cycle d of the converter. It is further assumed that the converter operates in continuous conduction mode (CCM), and without parasitic elements. The following expression shows the state-space bilinear model of the buck-boost converter [23]:

$$\begin{aligned}
\dot{\tilde{x}}(t) &= A\tilde{x}(t) + B_w\tilde{w}(t) + B_u(\tilde{x})\tilde{u}(t) \\
\tilde{z}(t) &= C_z\tilde{x}(t) + D_{zw}\tilde{w}(t) + D_{zu}\tilde{u}(t)
\end{aligned} \tag{1}$$

where $\tilde{x}(t)$, $\tilde{u}(t)$, $\tilde{w}(t)$ are the incremental averaged values of the state vector, input vector, and disturbance inputs around equilibrium values X, U and W, respectively. Besides, $\tilde{z}(t)$ represents the controlled output $\tilde{v}_o(t)$. Thus, based on the circuit of Figure 1, the expression (1) can be written as :

$$\tilde{x}(t) = \begin{bmatrix} \tilde{i}_L \\ \tilde{v}_C \end{bmatrix} \quad X = \begin{bmatrix} \frac{V_g D}{R D'^2} \\ -\frac{V_g D}{D'} \end{bmatrix} \quad \tilde{u}(t) = \begin{bmatrix} \tilde{d}(t) \end{bmatrix} \quad \tilde{w}(t) = \begin{bmatrix} \tilde{i}_o(t) \end{bmatrix} \quad \tilde{z}(t) = [\tilde{v}_C(t)]$$

$$A = \begin{bmatrix} 0 & \frac{D'}{L} \\ -\frac{D'}{C} & -\frac{1}{RC} \end{bmatrix} \quad B_u(\tilde{x}) = \begin{bmatrix} \frac{V_g}{D'L} - \frac{\tilde{v}_C(t)}{L} \\ \frac{V_g D}{(D'^2 R)C} + \frac{\tilde{i}_L(t)}{C} \end{bmatrix} \quad B_w = \begin{bmatrix} 0 \\ -\frac{1}{C} \end{bmatrix}$$

$$C_z = \begin{bmatrix} 0 & 1 \end{bmatrix} \quad D_{zw} = [0] \quad D_{zu} = [0]$$

$$\tag{2}$$

where $D' = 1 - D$ is the complementary steady-state duty cycle.

In order to ensure zero steady state error in the output voltage $v_o(t)$, a new state variable $x_3(t) = \int\left(v_o(t) - V_{ref}\right)$ has been introduced in the model (2). This integral function forces $v_C(t) \to V_{ref}$ when $t \to \infty$, where V_{ref} is the voltage reference.

Takagi-Sugeno Fuzzy representation allows to describe the dynamics of a non-linear system by means of a set of local linear models based on fuzzy rules, which are smoothly connected by membership functions. The rule set of a T-S fuzzy model is written as:

$$R_i : If \;\; \delta_1 \;\; is \;\; M_{i1} \;\; and \ldots and \;\; \delta_j \;\; is \;\; M_{ji} \;\; then$$

$$\begin{cases} \dot{x}(t) = A_i x(t) + B_{u_i} u(t) + B_{w_i} w(t) & i = 1, 2, \ldots r \\ \dot{z}(t) = C_{z_i} x(t) \end{cases} \tag{3}$$

where r is the number of the submodels, A_i are system matrices of the i-th linear submodel, B_{u_i} are input matrices, B_{w_i} are disturbance inputs matrices, C_{z_i} are controlled output matrices, $x(t)$ is the global state-space vector, $u(t)$ is the input vector, $w(t)$ is the disturbance input vector, $z(t)$ is the controlled output vector, M_{ji} are the fuzzy sets, and δ_j are the scheduling vector or premise variables [13]. On the other hand, $\eta_j(\delta_j)$ are the membership functions of the fuzzy sets M_{ji} and $h_i(\delta(t)) = \Pi_{j=1}^{n}\eta_j(\delta_j)$ the weight contribution of the rule. The final outputs of the fuzzy systems are obtained as the weighted sum of all the local contributing submodels, leading to:

$$\begin{aligned}
\dot{x}(t) &= \sum_{i=1}^{r} h_i(\delta(t))[A_i x(t) + B_{u_i} u(t) + B_{w_i} w(t)] \\
\dot{z}(t) &= \sum_{i=1}^{r} h_i(\delta(t))[C_{z_i} x(t)]
\end{aligned} \tag{4}$$

where, $0 \leq h_i(\delta(t)) \leq 1$, $\sum_{i=1}^{r} h_i(\delta(t)) = 1$, for $i = 1, 2, \ldots r$.

From the incremental bilinear model (2), the T-S fuzzy model of the buck-boost converter can be obtained [20,21]. With the selection of the scheduling variables $\delta(t)$, in most cases the state variables, the local linear models ($r = 2^j$) necessary for the construction of the total fuzzy model are obtained from the extreme values of $\delta(t)$. These extreme values will allow defining the rule base (R_i) of the fuzzy model, as well as the membership functions $\eta_j(\delta_j)$ necessary for the fuzzy weighting of the locally valid linear submodels associated to each R_i implication. Thus, the T-S fuzzy model approximation of the buck-boost converter can be constructed using the following steps:

1. Find the system scheduling variables $\delta(t)$:

$$\dot{\tilde{x}}(t) = A(\delta(t))\tilde{x}(t) + B_u(\delta(t))\tilde{d}(t) \tag{5}$$

$$A(\delta(t)) = A = \begin{bmatrix} 0 & \frac{D'}{L} \\ -\frac{D'}{C} & -\frac{1}{RC} \end{bmatrix} \qquad B_u(\delta(t)) = \begin{bmatrix} \frac{V_g}{D'L} - \frac{\delta_2(t)}{L} \\ \frac{V_g D}{D'^2 RC} + \frac{\delta_1(t)}{C} \end{bmatrix} \tag{6}$$

being $\delta(t) = [\delta_1(t) \quad \delta_2(t)] = \left[\tilde{i}_L(t) \quad \tilde{v}_o(t)\right]$.

2. Calculate $(r = 2^j)$ local linear models from extreme values of $\delta(t)$. For the ordered pairs: $(\delta_{1_{min}}, \delta_{2_{min}})$, $(\delta_{1_{max}}, \delta_{2_{min}})$, $(\delta_{1_{min}}, \delta_{2_{max}})$, $(\delta_{1_{max}}, \delta_{2_{max}})$ is obtained:

$$B_{u_1} = \begin{bmatrix} \frac{V_g}{D'L} - \frac{\delta_{2_{min}}}{L} \\ \frac{V_g D}{(D'^2 R)C} + \frac{\delta_{1_{min}}}{C} \end{bmatrix} \qquad B_{u_2} = \begin{bmatrix} \frac{V_g}{D'L} - \frac{\delta_{2_{min}}}{L} \\ \frac{V_g D}{(D'^2 R)C} + \frac{\delta_{1_{max}}}{C} \end{bmatrix}$$

$$B_{u_3} = \begin{bmatrix} \frac{V_g}{D'L} - \frac{\delta_{2_{max}}}{L} \\ \frac{V_g D}{(D'^2 R)C} + \frac{\delta_{1_{min}}}{C} \end{bmatrix} \qquad B_{u_4} = \begin{bmatrix} \frac{V_g}{D'L} - \frac{\delta_{2_{max}}}{L} \\ \frac{V_g D}{(D'^2 R)C} + \frac{\delta_{1_{max}}}{C} \end{bmatrix} \tag{7}$$

3. Design the membership functions. From the extreme values of $\delta(t)$, the membership functions are defined as follows:

$$\eta_{small}(\delta_1) = \frac{\delta_{1_{max}} - \delta_1}{\delta_{1_{max}} - \delta_{1_{min}}} \qquad \eta_{big}(\delta_1) = 1 - \eta_{small}(\delta_1)$$

$$\eta_{small}(\delta_2) = \frac{\delta_{2_{max}} - \delta_2}{\delta_{2_{max}} - \delta_{2_{min}}} \qquad \eta_{big}(\delta_2) = 1 - \eta_{small}(\delta_2) \tag{8}$$

4. Build up the rule-base R_i of the T-S fuzzy model. Fuzzy model of the buck-boost converter is defined by the following four rules:

$$\begin{aligned}
R_1: & \quad \text{If } \delta_1 \text{ is small and } \delta_2 \text{ is small} & \text{then} & \quad \dot{\tilde{x}}_1(t) = A_1\tilde{x}(t) + B_{u_1}\tilde{d}(t) \\
R_2: & \quad \text{If } \delta_1 \text{ is big and } \delta_2 \text{ is small} & \text{then} & \quad \dot{\tilde{x}}_2(t) = A_2\tilde{x}(t) + B_{u_2}\tilde{d}(t) \\
R_3: & \quad \text{If } \delta_1 \text{ is small and } \delta_2 \text{ is big} & \text{then} & \quad \dot{\tilde{x}}_3(t) = A_3\tilde{x}(t) + B_{u_3}\tilde{d}(t) \\
R_4: & \quad \text{If } \delta_1 \text{ is big and } \delta_2 \text{ is big} & \text{then} & \quad \dot{\tilde{x}}_4(t) = A_4\tilde{x}(t) + B_{u_4}\tilde{d}(t)
\end{aligned}$$

It is worth mentioning that in the literature on fuzzy controllers, membership functions with different shapes such as triangular, trapezoidal, Gaussian, among others, have been used. In this paper, for the sake of simplicity in practice, triangular membership functions were used.

Once the above steps are performed, the total fuzzy model of the converter can be expressed as:

$$\dot{\tilde{x}}(t) = \sum_{i=1}^{r} h_i\left(A_i\tilde{x}(t) + B_{u_i}\tilde{d}(t)\right) \tag{9}$$

Since $\sum_{i=1}^{r} h_i = 1$ and $A_i = A$, the fuzzy model (9) can be rewritten as:

$$\dot{\tilde{x}}(t) = A\tilde{x}(t) + \left(\sum_{i=1}^{r} h_i(\delta_1, \delta_2) B_{u_i}\right)\tilde{d}(t) \tag{10}$$

Therefore, the buck-boost converter fuzzy dynamics can be written as:

$$\dot{\tilde{x}}(t) = \begin{bmatrix} 0 & \frac{D'}{L} \\ -\frac{D'}{C} & -\frac{1}{RC} \end{bmatrix} \tilde{x}(t) + \left(h_1 \begin{bmatrix} \frac{V_g}{D'L} - \frac{v_{C_{min}}}{L} \\ \frac{V_g D}{(D'^2 R)C} + \frac{i_{L_{min}}}{C} \end{bmatrix} + h_2 \begin{bmatrix} \frac{V_g}{D'L} - \frac{v_{C_{min}}}{L} \\ \frac{V_g D}{(D'^2 R)C} + \frac{i_{L_{max}}}{C} \end{bmatrix} \right.$$

$$\left. + h_3 \begin{bmatrix} \frac{V_g}{D'L} - \frac{v_{C_{max}}}{L} \\ \frac{V_g D}{(D'^2 R)C} + \frac{i_{L_{min}}}{C} \end{bmatrix} + h_4 \begin{bmatrix} \frac{V_g}{D'L} - \frac{v_{C_{max}}}{L} \\ \frac{V_g D}{(D'^2 R)C} + \frac{i_{L_{max}}}{C} \end{bmatrix} \right) \tilde{d}(t) \quad (11)$$

where

$$\begin{aligned} h_1(\delta_1, \delta_2) &= \eta_{small}(\delta_1).\eta_{small}(\delta_2) & h_3(\delta_1, \delta_2) &= \eta_{small}(\delta_1).\eta_{big}(\delta_2) \\ h_2(\delta_1, \delta_2) &= \eta_{big}(\delta_1).\eta_{small}(\delta_2) & h_4(\delta_1, \delta_2) &= \eta_{big}(\delta_1).\eta_{big}(\delta_2) \end{aligned} \quad (12)$$

It should be noted that (9) represents exactly the model bilinear (2) in the polytopic region $\left[\delta_{1_{min}}, \delta_{1_{max}}\right] \times \left[\delta_{2_{min}}, \delta_{2_{max}}\right]$ which is shown in Figure 2.

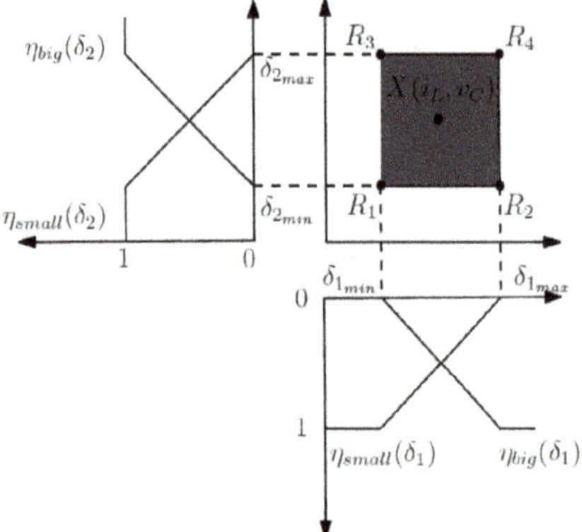

Figure 2. T-S Fuzzy representation of the buck-boost converter.

The model (11) will be used in section IV to find the T-S Fuzzy Control of the buck-boost converter. It is worth noting that the procedure can be used if there are more scheduling variables.

2.2. T-S Fuzzy Model of a Boost Converter

Figure 3 shows the schematic of a boost converter. As in the previous section, the converter operates in CCM and parasitic resistances in the inductor and the capacitor are sufficiently small to be neglected.

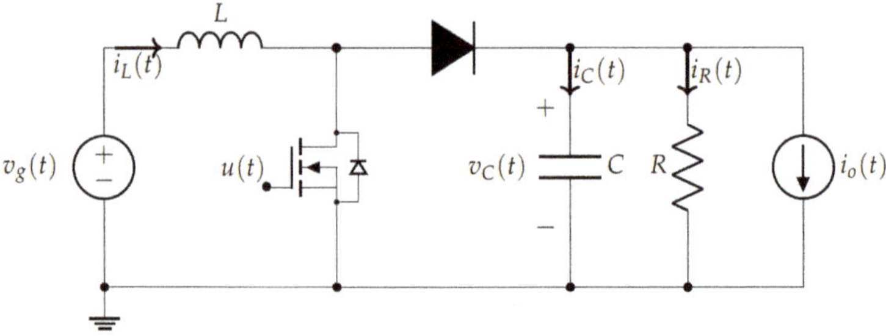

Figure 3. Schematic circuit of a boost converter.

The incremental bilinear model of the boost converter has the form presented in (1). The values of vectors and state space matrices are written as follows:

$$\tilde{x}(t) = \begin{bmatrix} \tilde{i}_L \\ \tilde{v}_C \end{bmatrix} \quad X = \begin{bmatrix} \frac{V_g}{RD'^2} \\ \frac{V_g}{D'} \end{bmatrix} \quad \tilde{u}(t) = \begin{bmatrix} \tilde{d}(t) \end{bmatrix} \quad \tilde{w}(t) = \begin{bmatrix} \tilde{i}_o(t) \end{bmatrix} \quad \tilde{z}(t) = \begin{bmatrix} \tilde{v}_C(t) \end{bmatrix}$$

$$A = \begin{bmatrix} 0 & -\frac{D'}{L} \\ \frac{D'}{C} & -\frac{1}{RC} \end{bmatrix} \quad B_u(\tilde{x}) = \begin{bmatrix} \frac{V_g}{D'L} + \frac{\tilde{v}_C(t)}{L} \\ -\frac{V_g}{D'^2RC} - \frac{i_L(t)}{C} \end{bmatrix} \quad B_w = \begin{bmatrix} 0 \\ -\frac{1}{C} \end{bmatrix}$$

$$C_z = \begin{bmatrix} 0 & 1 \end{bmatrix} \quad D_{zw} = [0] \quad D_{zu} = [0]$$

$$(13)$$

Applying the methodology of the previous section, the T-S fuzzy model of the boost converter can be expressed as:

$$\tilde{x}(t) = \begin{bmatrix} 0 & -\frac{D'}{L} \\ \frac{D'}{C} & -\frac{1}{RC} \end{bmatrix} \tilde{x}(t) + \left(h_1 \begin{bmatrix} \frac{V_g}{D'L} + \frac{v_{C_{min}}}{L} \\ -\frac{V_g}{(D'^2R)C} - \frac{i_{L_{min}}}{C} \end{bmatrix} + h_2 \begin{bmatrix} \frac{V_g}{D'L} + \frac{v_{C_{min}}}{L} \\ -\frac{V_g}{(D'^2R)C} - \frac{i_{L_{max}}}{C} \end{bmatrix} \right.$$
$$\left. + h_3 \begin{bmatrix} \frac{V_g}{D'L} + \frac{v_{C_{max}}}{L} \\ -\frac{V_g}{(D'^2R)C} - \frac{i_{L_{min}}}{C} \end{bmatrix} + h_4 \begin{bmatrix} \frac{V_g}{D'L} + \frac{v_{C_{max}}}{L} \\ -\frac{V_g}{(D'^2R)C} - \frac{i_{L_{max}}}{C} \end{bmatrix} \right) \tilde{d}(t) \quad (14)$$

where the weight contribution of each fuzzy rule $\{h_1, \ldots, h_4\}$, have the same behavior as for the buck-boost converter case (12).

In the next section, the proposed LMI-fuzzy control strategy is explained. This control law consists of a weighted sum of the feedback gains of each submodel, which has constraints such as: perturbation rejection level, decay rate of state variables and control effort.

3. LMI Performance Requeriments for Fuzzy Controllers

The design procedure of the controller is based on the Parallel Distributed Compensation technique (PDC), which is used to design state feedback controllers based on T-S fuzzy models [13]. This metholodology consists in associating each control rule with the corresponding rule of the fuzzy model, as follows:

$$R_i : \textbf{If } \delta_1 \text{ is } M_{i1} \text{ and} \ldots \text{and } \delta_j \text{ is } M_{ji} \textbf{ then } \quad u(t) = -F_i x(t) \quad i = 1, \ldots, r \quad (15)$$

where F_i are lineal feedback gain vectors associated with each rule. Then, the output controller is deduced as:

$$u(t) = -\sum_{i=1}^{r} h_i F_i x(t) \quad (16)$$

Substituting the control law (16) in the fuzzy model (4), the closed loop system dynamics is given by:

$$\dot{x}(t) = \sum_{i=1}^{r} \sum_{j=1}^{r} h_i(\delta) h_j(\delta) [A_i + B_i F_j] x(t) \quad (17)$$

In order to find the feedback gain vectors properly F_i in an operating range, Lyapunov stability and performance constraints in form of LMIs are imposed. In this way, these performance constraints, taken from [13], are expressed by the following theorems.

Theorem 1. *The system defined by (4) is quadratically stable for some feedback gains F_i and $\frac{\|z\|_2}{\|w\|_2} < \gamma$ if there is a common positive definite matrix W and Y_i such that [13]:*

$$\begin{bmatrix} \left(\begin{array}{c} \frac{1}{2}\{A_i W + W A_i^T + A_j W + W A_j^T + \\ - B_i Y_j - Y_j^T B_i^T - B_j Y_i - Y_i^T B_j^T \} \end{array} \right) & \frac{B_{w_i}+B_{w_j}}{2} & -W\frac{\left(C_{z_i}+C_{z_j}\right)^T}{2} \\ \frac{\left(B_{w_i}+B_{w_j}\right)^T}{2} & -\gamma I & 0 \\ -\frac{\left(C_{z_i}+C_{z_j}\right)}{2} W & 0 & -I \end{bmatrix} < 0 \qquad (18)$$

where I is the identity matrix, $W = P^{-1}$ and $F_i = Y_i W^{-1}$.

Theorem 2. *The eigenvalues of $\left(A_i - B_i F_j\right)$ in each linear fuzzy system are inside the region $S(\alpha)$ (see Figure 4) if there is a common positive definite matrix W such that [13]:*

$$A_i W + W A_i^T - B_i Y_i - Y_i^T B_i^T + 2\alpha W < 0, \qquad\qquad i = 1,\ldots r$$

$$A_i W + W A_i^T + A_j W + W A_j^T - B_i Y_j - Y_j^T B_i^T - B_j Y_i - Y_i^T B_j^T + 4\alpha W \leq 0, \quad i < j \leq r$$

$$(19)$$

being $Y_i = F_i W$ so that for $W > 0$, it is had $F_i = Y_i W^{-1}$.

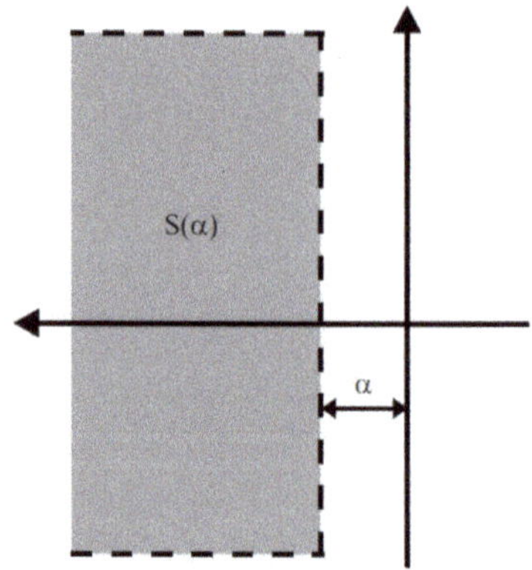

Figure 4. Pole location region $S(\alpha)$.

Theorem 3. *Assume that initial condition $x(0)$ is known. The constraint $\|d(t)\|_2 \leq \mu$ is enforced at all times $t \geq 0$ if the LMIs (20) hold [13]:*

$$\begin{bmatrix} 1 & x(0)^T \\ x(0) & W \end{bmatrix} \geq 0,$$

$$\begin{bmatrix} W & Y_i^T \\ Y_i & \mu^2 I \end{bmatrix} \geq 0 \qquad (20)$$

where $W = P^{-1}$ and $Y_i = F_i W$.

A detailed proof of the Theorems 1, 2 and 3 are shown in [13]. Thus, from the previous LMIs, the procedure proposed is to find the minimum norm H_∞ (γ) between the disturbance input and the regulated output, ensuring at the same time stability in a wide domain of operation. The synthesis of the LMI-Fuzzy control can be performed using the following optimization algorithm:

$$\begin{aligned} &min \; \gamma \quad subject \; to \\ &\mathbf{W}, \mathbf{Y}_i \\ &(18), (19), \; and \; (20) \quad \forall \; i = 1, \ldots, r \end{aligned} \tag{21}$$

The solution of this optimization program with its corresponding LMIs will provide the set of feedback gains $\mathbf{F}_{LMI-Fuzzy} = \{\mathbf{F}_1, \ldots, \mathbf{F}_r\}$. The solution of this algorithm can be readily solved by standard interior-point methods using Matlab [24].

4. Simulations and Experimental Results

This section presents two examples of LMI-Fuzzy control design applied to the non-minimum phase dc-dc converters for the output voltage regulation problem. In the first case, a fuzzy control is applied to a buck-boost converter, taking into account the T-S fuzzy model described in Section 2.1. The second example proposes a fuzzy control applied to a boost converter with its corresponding T-S fuzzy model Section 2.2. Both control examples ensure fulfilment of the LMI restrictions on decay rate and control effort, which correspond to the LMIs (19) and (20), optimizing the rejection of disturbances in the load current (18), by applying the algorithm (21). In addition, this section shows some simulations in PSIM of both examples, where the proposed approach is compared with non-fuzzy LMI robust controller. Finally, for the case of the step-up converter, the validity of the design procedure is demonstrated through experimental results.

4.1. LMI-Fuzzy Control of a Buck-Boost Converter

In this first example, as mentioned above, the design of an LMI-Fuzzy control law is shown as an alternative to the voltage regulation of the buck-boost converter, based on the fuzzy model (11), whose set of parameters is shown in Table 1. The values of the state variables in steady state, according to the expression (2), corresponds to $[I_L, V_C] = [4.8\,\text{A}, -24\,\text{V}]$. Thus, the simulation prototype of the converter is considered for processes smaller than 60 W, taking into account a load resistance $R = 10\,\Omega$.

Table 1. Buck-Boost converter parameters.

V_g	$v_o(V_{ref})$	L	C	R	R_2	$[\tilde{i}_{min} \times \tilde{i}_{max}]$	$[\tilde{v}_{min} \times \tilde{v}_{max}]$	D'	T_s
24 V	−24 V	200 μH	200 μF	10 Ω	20 Ω	$[-30, \; 20]$ A	$[0, \; 50]$ V	0.5	10 μs

The goal of control synthesis is to find a vector of feedback gains $\mathbf{F}_{LMI-Fuzzy}$ such that the norm H_∞ (γ) is minimized, satisfying the constraints on the decay rate and the control effort for the four linear submodels that build the T-S fuzzy model (11). The chosen controller parameter values (α, μ) are specified in Table 2.

Table 2. Controller Parameters.

α	μ
450 s^{-1}	7

It should be noted that the decay rate α that has been chosen corresponds to a maximum time of establishment of $4 * (1/450)$ s, while the threshold limit value for the saturation of the control signal, corresponds to $\mu = 7$. In this way, solving the optimization algorithm (21) through MATLAB's LMI toolbox [24], the fuzzy state feedback gains are obtained $\mathbf{F}_{LMI-Fuzzy} = \{\mathbf{F}_1, \ldots, \mathbf{F}_4\}$:

$$\mathbf{F}_1 = \begin{bmatrix} -0.1094 & 0.1537 & -165.2341 \end{bmatrix} \quad \mathbf{F}_2 = \begin{bmatrix} -0.1231 & 0.2248 & -196.2015 \end{bmatrix}$$
$$\mathbf{F}_3 = \begin{bmatrix} -0.0802 & 0.1041 & -121.4163 \end{bmatrix} \quad \mathbf{F}_4 = \begin{bmatrix} -0.1230 & 0.2254 & -201.5697 \end{bmatrix}$$

This controller ensures a H_∞ gain of the output voltage with respect to the output current disturbance of $\gamma = 1.3$, which is equivalent to 2.2789 dB. In a general case, the corresponding fuzzy law produced by the duty cycle will be given by:

$$\tilde{d}(t) = [h_1\mathbf{F}_1 + h_2\mathbf{F}_2 + h_3\mathbf{F}_3 + h_4\mathbf{F}_4]\tilde{x}(t) \tag{22}$$

As mentioned at the beginning of the section, in order to evaluate the operation and robustness of the proposed methodology, the results are compared with a LMI robust linear control law, applying the approach proposed in [9]. For this reason, the vector of feedback gains $\mathbf{F}_{LMI}$ that is obtained corresponds to:

$$\mathbf{F}_{LMI} = [-0.6292 \quad 0.5371 \quad -525.2079]$$

The H_∞ gain of the output voltage with respect to the output current disturbance of this controller is $\gamma = 2.2$, corresponding to 6.8485 dB. For this case, the generated equivalent law is given by:

$$\tilde{d}(t) = \mathbf{F}_{LMI}\tilde{x}(t) \tag{23}$$

Through the LMI-Fuzzy control schematic diagram for the converter buck-boost from Figure 5, some simulations of the dynamic behavior of the converter in the presence of changes in the load and in the input voltage were carried out in PSIM inside and outside the nominal conditions, taking into account the values in Tables 1 and 2.

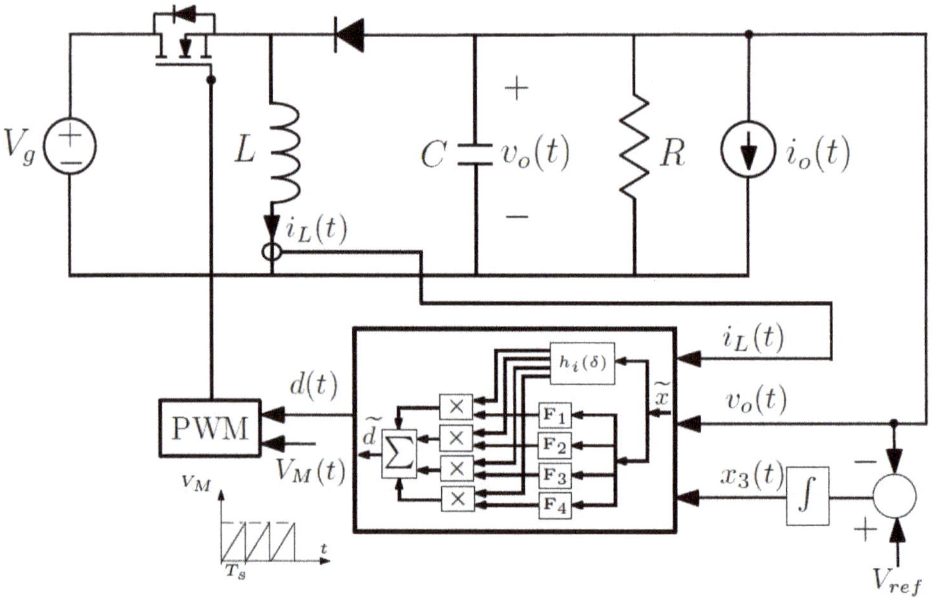

Figure 5. Circuital diagram of a buck-boost converter with the LMI-Fuzzy control.

Figure 6a depicts waveforms of the output voltage v_o of the buck-boost converter with LMI-Fuzzy control, under nominal conditions and in the face of changes in load current of

2 A. It can be seen that the controller regulates the voltage smoothly at -24 V after a short transient period, equivalent to a time constant of 4 ms ($\alpha = 1000$), which is a value that is above the minimum guaranteed decay rate that is ($\alpha = 450$). In Figure 6b, with the same previous conditions, the output voltage response is shown with the LMI robust control proposed in [9]. It is worth noting that the voltage response presents a time constant of approximately 5 ms, equivalent to a decay rate of $\alpha = 800$, as expected with the minimum guaranteed decay rate. However, comparing the two previous results, it can be highlighted that the LMI-Fuzzy control law presents better dynamic behavior than the LMI robust control law, since it shows a better decay rate and disturbances rejection.

On the other hand, in Figure 7 the dynamic behavior of the buck-boost converter is described under input voltage variations with the laws LMI-Fuzzy and LMI robust. Namely, an input-voltage step from 24 V to 22 V is applied at $t = 4$ ms and returned at $t = 24$ ms. Also, it can be observed that for both controllers the output voltage response presents decay rates greater than the guaranteed minimum. ($\alpha = 450$). As in the previous case, the fuzzy control law presents better dynamic behavior, both in decay rate and in disturbances rejection.

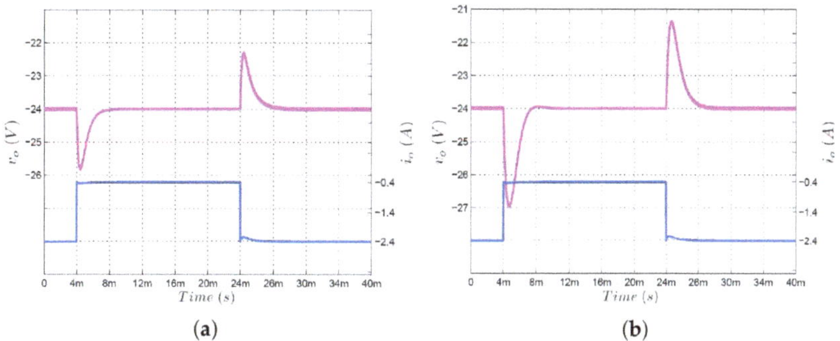

Figure 6. Simulated response of the buck-boost converter under output current transients of 2 A with the LMI-Fuzzy controller ($\mathbf{F}_{LMI-Fuzzy}$) and LMI robust controller ($\mathbf{F}_{LMI}$). (**a**) Voltage response v_o with $\mathbf{F}_{LMI-Fuzzy}$. (**b**) Voltage response v_o with $\mathbf{F}_{LMI}$ [9].

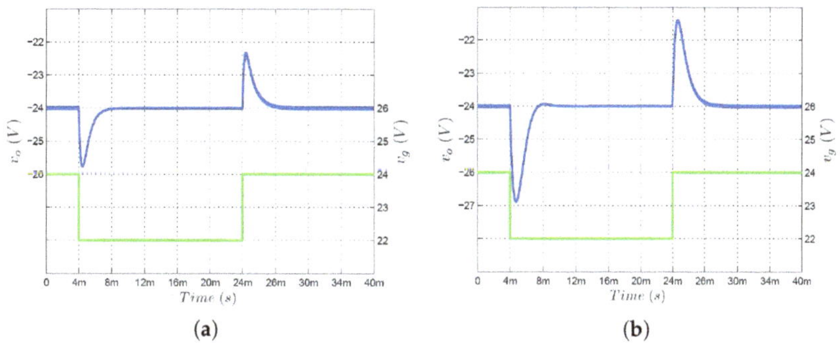

Figure 7. Simulated response of the buck-boost converter under input voltage transients of 2 V with the LMI-Fuzzy controller ($\mathbf{F}_{LMI-Fuzzy}$) and LMI robust controller ($\mathbf{F}_{LMI}$). (**a**) Voltage response v_o with $\mathbf{F}_{LMI-Fuzzy}$. (**b**) Voltage response v_o with $\mathbf{F}_{LMI}$ [9].

In the next subsection, the design of a LMI-Fuzzy control will be presented for the case of a boost converter.

4.2. LMI-Fuzzy Control of a Boost Converter

As in the previous subsection, the regulation of the output voltage for a step-up converter is presented, taking into account its corresponding fuzzy model (12). This design consists of solving the optimization algorithm (21) for the set of parameters shown in Table 3, where the nominal load of the converter is equal to 10 Ω and the complementary duty cycle in steady state is equal to 0.5.

Table 3. Boost converter parameters.

V_g	$v_o(V_{ref})$	L	C	R	R_2	$[\tilde{i}_{min} \times \tilde{i}_{max}]$	$[\tilde{v}_{min} \times \tilde{v}_{max}]$	D'	T_s
12 V	24 V	88 µH	200 µF	10 Ω	20 Ω	[0, 50] A	[20, 30] V	0.5	10 µs

As it has been established, the control synthesis procedure consists of finding a vector of feedback gains $F_{LMI-Fuzzy}$ such that the parameter γ is minimized in the LMI (20), while the constraints on the decay rate (α) and the control effort are satisfied for the four linear submodels that build the T-S fuzzy model (12). The values of the control parameters, i.e., (α, μ), are the same used for the case of the buck-boost converter. In order to demonstrate the advantage of the fuzzy control, as in the previous example, a comparison is made with the results of the LMI robust linear control law proposed in [9]. Therefore, for the set of parameters in Tables 2 and 3, the feedback gain vectors $\mathbf{F}_{LMI-Fuzzy}$ and $\mathbf{F}_{LMI}$ are obtained. In this way, for the case of LMI-Fuzzy control, the fuzzy state feedback gains correspond to:

$$\mathbf{F}_1 = \begin{bmatrix} -0.6 & -0.982 & 1229.7 \end{bmatrix} \quad \mathbf{F}_2 = \begin{bmatrix} -0.7 & -1.272 & 1498.7 \end{bmatrix}$$
$$\mathbf{F}_3 = \begin{bmatrix} -0.97 & -1.67 & 2053.7 \end{bmatrix} \quad \mathbf{F}_4 = \begin{bmatrix} -1.01 & -1.824 & 2143.6 \end{bmatrix} \tag{24}$$

This controller ensures a H_∞ gain of the output voltage with respect to the output current disturbance of $\gamma = 1.0145$ (0.1250 dB). Furthermore, it is shown that the matrix $\mathbf{P}$ is positive definite, ensuring that the asymptotic stability of the converter is fulfilled, as shown in (25).

$$\mathbf{P} = \mathbf{W}^{-1} = \begin{bmatrix} 4.292 \times 10^{-4} & 7.699 \times 10^{-4} & -0.919 \\ 7.699 \times 10^{-4} & 1.4094 \times 10^{-4} & -1.668 \\ -0.9193 & -1.668 & 2223.954 \end{bmatrix} \quad \mathbb{R}(\lambda(\mathbf{P})) = \begin{bmatrix} 6.663 \times 10^{-6} \\ 1.9979 \times 10^{-4} \\ 2223.9 \end{bmatrix} > 0 \tag{25}$$

The LMI robust feedback gain vector $\mathbf{F}_{LMI}$ obtained corresponds to:

$$\mathbf{F}_{LMI} = \begin{bmatrix} -0.5555 & -0.6090 & 743.8420 \end{bmatrix} \tag{26}$$

This controller ensures a minimum level of disturbance rejection equivalent to the inverse of $\gamma = 2.31$ (7.2 dB).

In order to verify the dynamic behavior of the boost converter under the laws LMI robust (26) and LMI-Fuzzy (24), some numerical simulations were performed via PSIM.

Figure 8a depicts waveforms of the output current i_o and the regulated voltage v_o of the converter, controlled by the feedback gains (24). The bottom waveform shows a change from 2.4 A to 3.6 of the output current i_o at $t = 4$ ms, and the opposite transition at $t = 24$ ms. As expected, the settling time is 8.9 ms below, which corresponds to the minimum decay rate set in the Table 2. Now, the same current disturbance i_o is applied to the boost converter controlled by the feedback gain (26), where it can be seen from the waveform that the settling time is once again within the chosen design limit. As in the case of the buck-boost converter, it is worth noting that the results of the LMI-Fuzzy control show better dynamic behaviour than the LMI robust control, both in terms of settling time and disturbance rejection. Figure 9a,b illustrate the responses of the output voltage v_o to an input voltage variation, for both the law of LMI-Fuzzy control and the law of LMI robust control, respectively. The input voltage V_g changes from 12 V to 10 V at $t = 4$ ms and returns to 12 V at $t = 24$ ms. Again, the waveforms have a settling time below the minimum set decay rate. Also, it can be observed that the disturbance rejection is better in the case of LMI-Fuzzy control than in the LMI robust control.

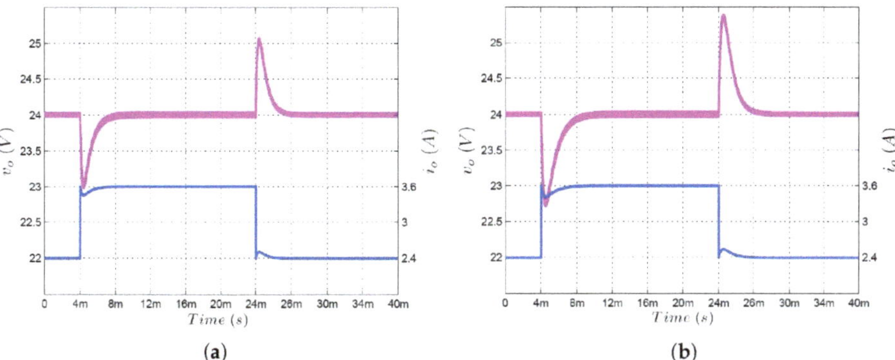

Figure 8. Simulated response of the boost converter under output current transients of 2 A with the LMI-Fuzzy controller ($\mathbf{F}_{LMI-Fuzzy}$) and LMI robust Controller ($\mathbf{F}_{LMI}$). (**a**) Voltage response v_o with $\mathbf{F}_{LMI-Fuzzy}$. (**b**) Voltage response v_o with $\mathbf{F}_{LMI}$ [9].

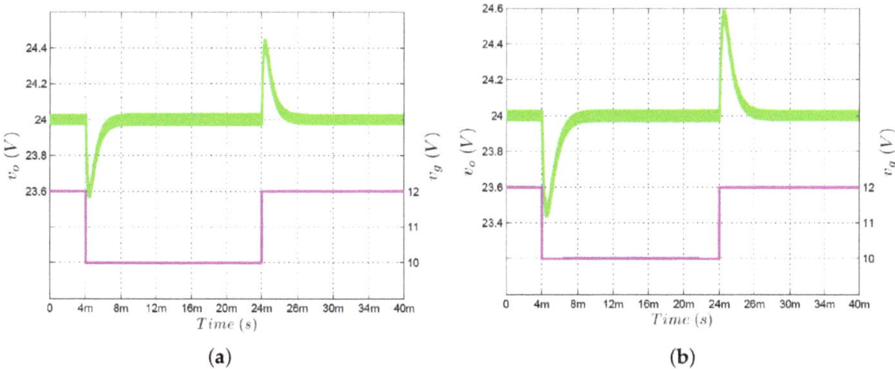

Figure 9. Simulated response of the boost converter under input voltage transients of 2 V with the LMI-Fuzzy controller ($\mathbf{F}_{LMI-Fuzzy}$) and LMI robust Controller ($\mathbf{F}_{LMI}$). (**a**) Voltage response v_o with $\mathbf{F}_{LMI-Fuzzy}$. (**b**) Voltage response v_o with $\mathbf{F}_{LMI}$ [9].

In order to verify the theoretical predictions of the LMI-Fuzzy control and LMI robust control, an experimental prototype of the boost converter has been implemented whose characteristics are shown in Table 3. Figure 10 shows the circuit diagram of the prototype with the feedback structure that was used for the LMI-Fuzzy control, where it can be seen that the measurement of the current i_L is carried out through the shunt resistance R_s, with a value of 25 mΩ; also a differential amplifier INA139 was used.

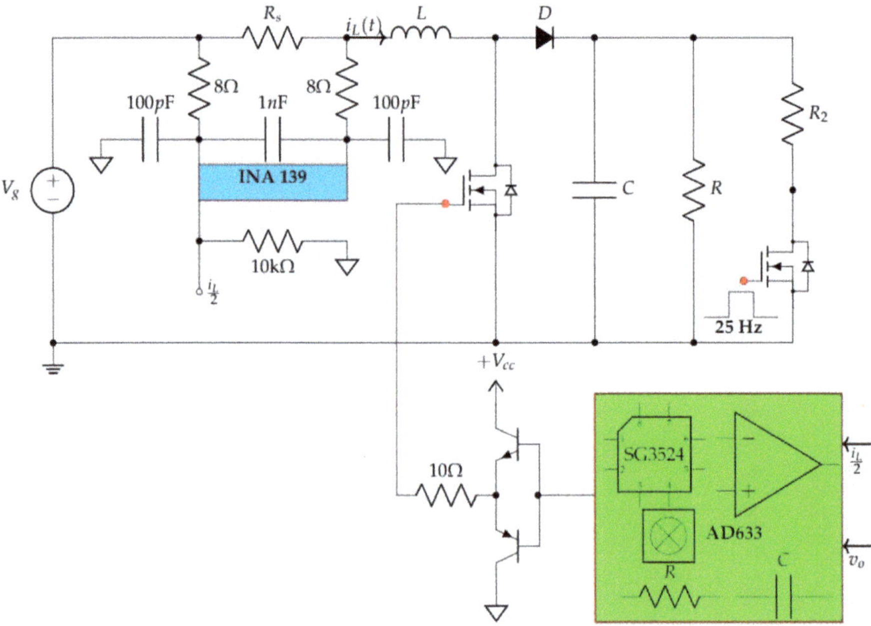

Figure 10. Circuit diagram of a boost converter with the proposed LMI-Fuzzy control.

The power stage of boost converter used for the application of both controllers is illustrated in Figure 11. This has been implemented using an IRFP150NPBF mosfet, and a MBR745 Schottky diode, which are activated by a PWM signal generated by the UC3524 regulator through a totem-pole configuration driver. Note that the output variables are: the current measured in the inductor $\frac{i_L}{2}(t)$ and the output voltage of the converter $v_o(t)$, while the inputs correspond to the duty cycle signal $d(t)$ and the supply signal V_{cc}. On the other hand, the implementation of the LMI-Fuzzy control is illustrated in Figure 12a,b. Both figures show the top sides of the printed circuit boards used in the tests. The Figure 13a,b illustrate the detailed circuits of the printed circuit boards of the control law $\mathbf{F}_{LMI-Fuzzy}$ (24). The circuit diagram in Figure 13a, is responsible for compute the membership functions (8) and the normalized weights of contribution of the rules (12), while the circuit diagram in Figure 13b calculates the linear combination (22) named total T-S fuzzy controller . Both control laws, i.e., $\mathbf{F}_{LMI-Fuzzy}$ and $\mathbf{F}_{LMI}$, were implemented with OPA4131 operational amplifiers. It is worth noting that the implementation of the analog LMI-Fuzzy controller is much more complex, since it requires more operational amplifiers and AD633 multipliers.

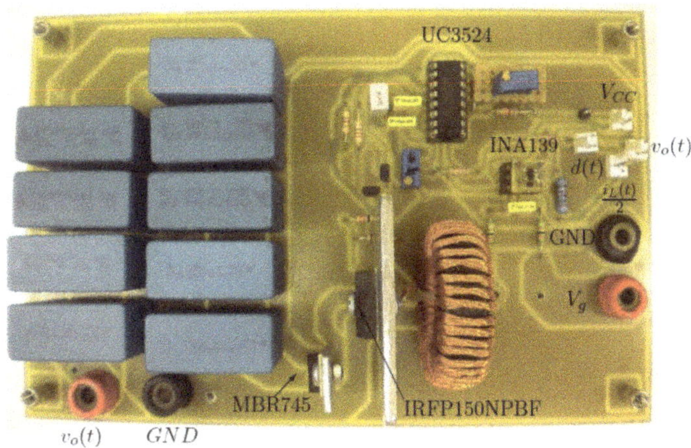

Figure 11. Boost converter prototype.

(**a**)	(**b**)

Figure 12. Practical LMI-Fuzzy control implementation. (**a**) Printed circuit board for the calculate membership functions and normalized weights. (**b**) Printed circuit board for the calculate total T-S fuzzy controller.

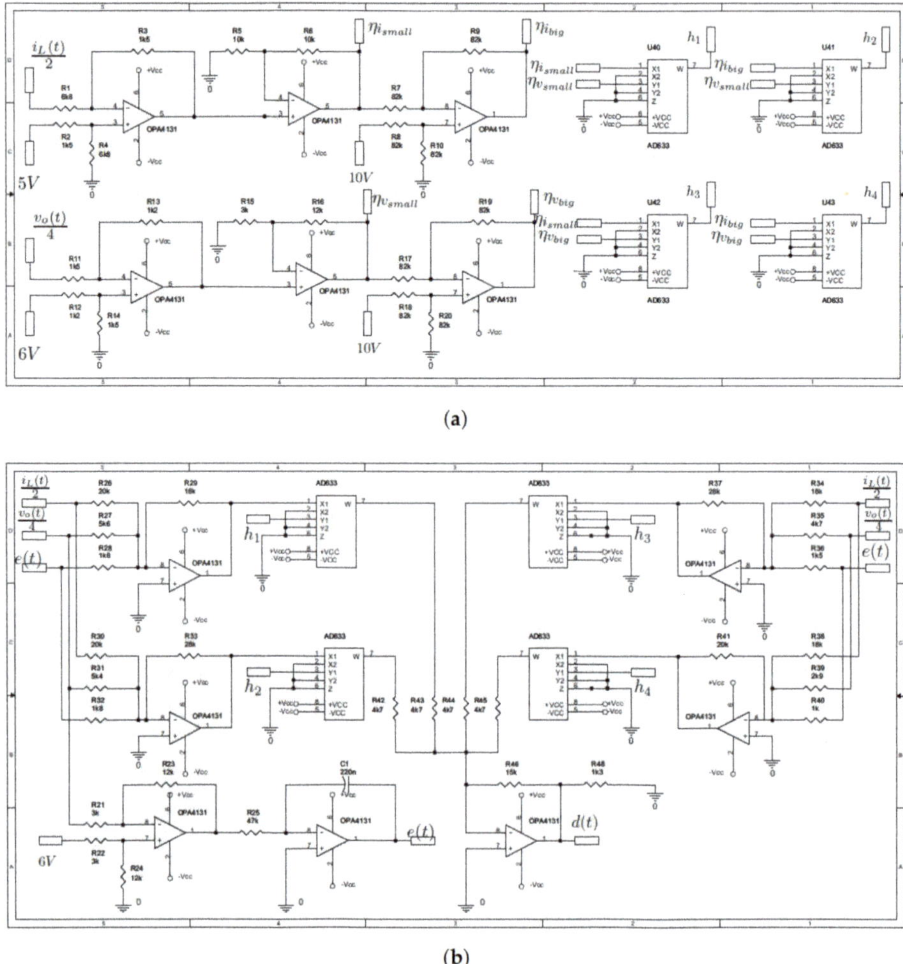

(a)

(b)

Figure 13. Detailed circuit implementation of the LMI-Fuzzy Control. (**a**) Circuit diagram for the calculate membership functions and normalized weights. (**b**) Circuit diagram for the calculate total T-S fuzzy controller.

Figure 14a,b show the transient responses of the output voltage in the presence of a 1.2 A load current disturbance, for the controllers (24) and (26), respectively. These experimental results and the previous simulated waveforms of Figure 8a,b are in very good agreement. The load current changes in the prototype were carried out by means of a voltage-controlled switch, such as shown in the circuit diagram of the converter in Figure 10. Furthermore, It was verified the response of the controllers $\mathbf{F}_{LMI-Fuzzy}$ and $\mathbf{F}_{LMI}$ to a supply voltage change. The experimental result, shown in Figure 15 accurately verifies the simulations shown in Figure 9.

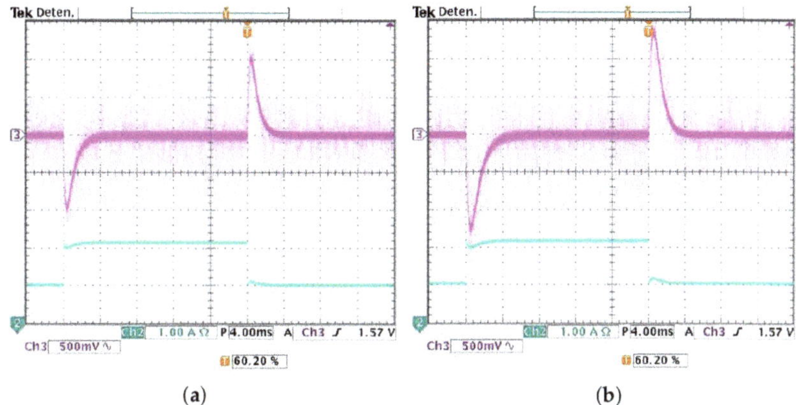

(a)   (b)

Figure 14. Experimental response of the boost converter under input voltage transients of 2 V with the LMI-Fuzzy controller ($\mathbf{F}_{LMI-Fuzzy}$) and LMI robust Controller ($\mathbf{F}_{LMI}$). (**a**) Voltage response v_o with $\mathbf{F}_{LMI-Fuzzy}$. (**b**) Voltage response v_o with $\mathbf{F}_{LMI}$ [9].

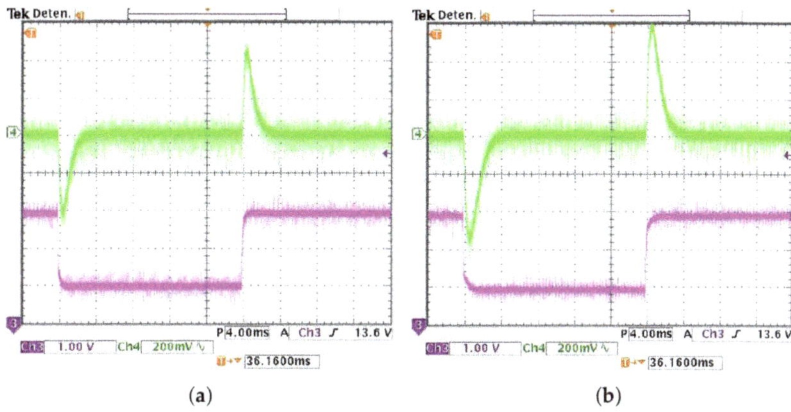

(a)   (b)

Figure 15. Experimental response of the boost converter under input voltage transients of 2 V with the LMI-Fuzzy controller ($\mathbf{F}_{LMI-Fuzzy}$) and LMI robust Controller ($\mathbf{F}_{LMI}$). (**a**) Voltage responde v_o with $\mathbf{F}_{LMI-Fuzzy}$. (**b**) Voltage response v_o with $\mathbf{F}_{LMI}$ [9].

5. Conclusions

In this paper a T-S fuzzy control approach based on LMIs has been presented for the output voltage regulation of non-minimum phase converters. The application of this approach focuses on the building of a T-S fuzzy model based on the bilinear nature of the converters, which is key to the design of fuzzy controllers. The control synthesis ensures fulfillment with constraints such as: decay rate of the state variables and the control effort while guaranteeing a minimum level of attenuation between output-current disturbance and the regulated output voltage. The application of the methodology is explained in detail by means of two design examples for the regulation of the basic buck-boost and boost converters. In the case of the boost converter, an experimental prototype was implemented to corroborate the theoretical predictions developed. In addition, in order to evaluate the performance of the proposed methodology, a comparison with a LMI non-fuzzy control a comparison with a non-fuzzy LMI control was performed using the approach proposed in [9], where the effectiveness of the LMI-Fuzzy control was proven, despite its complex implementation. The main contribution of this paper focuses on the experimental verification of the proposed design, which validates the theoretical predictions, in contrast to other works where only simulation results are described.

Author Contributions: All authors contributed equally to all the stages of the research process and the preparation of the manuscript. All authors have read and agreed to the published version of the manuscript.

Funding: This research was developed with the partial support of the Universidad Santo Tomás (Colombia) under contract proyecto FODEIN 2021 2154503, and the support of SERC Chile (CON-ICYT/FONDAP/15110019). Leonel Paredes-Madrid would like to acknowledge Antonio Nariño University for funding through grant PI/UAN-2021-686GIBIO.

Institutional Review Board Statement: The development of this work did not involve living species. The development of this research was carried out following the ethical standards and procedures defined by the authors' Institutions.

Informed Consent Statement: The development of this work did not require the collection of data from individuals or institutions.

Data Availability Statement: We claim that the experimental data reported in this manuscript is trustworthy.

Conflicts of Interest: The authors declare no conflict of interest.

References

1. Erickson, R.W.; Macksimovic, D. *Fundamental of Power Electronics*; Springer: Cham, Switzerland, 2001.
2. Calvente, J.; Martinez-Salamero, L.; Garces, P.; Romero, A. Zero dynamics-based design of damping networks for switching converters. *IEEE Trans. Aerosp. Electron. Syst.* **2003**, *39*, 1292–1303. doi:10.1109/TAES.2003.1261129.
3. Ramírez-Murillo, H.; Restrepo, C.; Calvente, J.; Romero, A.; Giral, R. Energy Management of a Fuel-Cell Serial–Parallel Hybrid System. *IEEE Trans. Ind. Electron.* **2015**, *62*, 5227–5235. doi:10.1109/TIE.2015.2395386.
4. Vidal-Idiarte, E.; Martinez-Salamero, L.; Valderrama-Blavi, H.; Guinjoan, F.; Maixe, J. Analysis and design of H_∞ control of nonminimum phase-switching converters. *IEEE Trans. Circuits Syst. Fundam. Theory Appl.* **2003**, *50*, 1316–1323. doi:10.1109/TCSI.2003.816337.
5. Houpis, C.H.; Rasmussen, S.J.; Garcia-Sanz, M. *Quantitative Feedback Theory: Fundamentals and Applications*, 2nd ed.; CRC Press: New York, NY, USA, 2005.
6. Olalla, C.; Leyva, R.; El Aroudi, A.; Garcés, P. QFT robust control of current-mode converters: Application to power conditioning regulators. *Int. J. Electron.* **2009**, *96*, 503–520. doi:10.1080/00207210802641239.
7. Montagner, V.F.; Peres, L.D. H_∞ control with pole location for a DC-DC converter with a switched load. In Proceedings of the 2003 IEEE International Symposium on Industrial Electronics (Cat. No.03TH8692), Rio de Janeiro, Brazil, 9–11 June 2003; Volume 1, pp. 550–555. doi:10.1109/ISIE.2003.1267310.
8. Montagner, V.F.; Oliveira, R.; Leite, V.; Peres, L.D. LMI approach for H_∞ linear parameter-varying state feedback control. *IEE Proc. Control. Theory Appl.* **2005**, *152*, 195–201. doi:10.1049/ip-cta:20045117.
9. Olalla, C.; Leyva, R.; El Aroudi, A.; Garcés, P.; Queinnec, I. LMI robust control design for boost PWM converters. *IET Power Electron.* **2010**, *3*, 75–85. doi:10.1049/iet-pel.2008.0271.
10. Olalla, C.; Queinnec, I.; Leyva, R.; El Aroudi, A. Robust optimal control of bilinear DC–DC converters. *Control Eng. Pract.* **2011**, *19*, 668–699. doi:10.1080/00207210802641239.
11. Olalla, C.; Queinnec, I.; Leyva, R.; El Aroudi, A. Optimal State-Feedback Control of Bilinear DC–DC Converters With Guaranteed Regions of Stability. *IEEE Trans. Ind. Electron.* **2012**, *59*, 3868–3880. doi:10.1109/TIE.2011.2162713.
12. Olalla, C.; Leyva, R.; Queinnec, I.; Maksimovic, D. Robust Gain-Scheduled Control of Switched-Mode DC–DC Converters. *IEEE Trans. Power Electron.* **2012**, *27*, 3006–3019. doi:10.1109/TPEL.2011.2178271.
13. Tanaka, K.; Wang, H. *Fuzzy Control Systems Design and Analysis: A Linear Matrix Inequality Approach*, 1st ed.; John Wiley & Sons, Inc.: New York, NY, USA , 2001.
14. Korba, P.; Babuska, R.; Verbruggen, H.B.; Frank, P.M. Fuzzy gain scheduling: Controller and observer design based on Lyapunov method and convex optimization. *IEEE Trans. Fuzzy Syst.* **2003**, *11*, 285–298. doi:10.1109/TFUZZ.2003.812680.
15. Hong, S.; Nam, Y. Stable fuzzy control system design with pole-placement constraint: An LMI approach. *Comput. Ind.* **2003**, *51*, 1–11. doi:10.1016/S0166-3615(03)00057-5.
16. Soliman, M.; Elshafei, A.; Bendary, F.; Mansour, W. LMI static output-feedback design of fuzzy power system stabilizers. *Expert Syst. Appl.* **2009**, *36*, 6817–6825. doi:10.1016/j.eswa.2008.08.018.
17. Lian, K.Y.; Liou, J.J.; Huang, C.Y. LMI-based Integral fuzzy control of DC-DC converters. *IEEE Trans. Fuzzy Syst.* **2006**, *14*, 71–80. doi:10.1109/TFUZZ.2005.861610.

18. Lan, H.K.; Tan, S. Stability analysis of fuzzy-model-based control systems: application on regulation of switching dc-dc converter. *IET Control Theory Appl.* **2009**, *3*, 1093–1106. doi:10.1049/iet-cta.2008.0168.
19. Saifia, D.; Chadli, M.; Labiod, S.; Karimi, H. H_∞ fuzzy control of DC-DC converters with input constraint. *Math. Probl. Eng.* **2012**, *2012*, 1–18. doi:10.1155/2012/973082.
20. Torres-Pinzón, C.A.; Leyva, R. Fuzzy control in DC-DC converters: An LMI approach. In Proceedings of the 2009 35th Annual Conference of IEEE Industrial Electronics, Porto, Portugal, 3–5 November 2009; pp. 510–515. doi:10.1109/IECON.2009.5414974.
21. Torres-Pinzon, C.A.; Leyva, R. MATLAB: A Systems Tool for Design of Fuzzy LMI Controller in DC-DC Converters. In *MATLAB*; Ionescu, C.M., Ed.; IntechOpen: Rijeka, Croatian, 2011; Chapter 13. doi:10.5772/23856.
22. Tanaka, T.; Sugeno, M. Stability analysis and design of fuzzy control systems. *Fuzzy Sets Syst.* **1992**, *45*, 135–156. doi:10.1016/0165-0114(92)90113-I.
23. Torres-Pinzon, C.; Giral, R.; Leyva, R. LMI-Based Robust Controllers for DC-DC Cascade Boost Converters. *J. Power Electron.* **2012**, *12*, 538–547. doi:10.6113/JPE.2012.12.4.538.
24. Gahinet, P.; Nemirovski, A.; Laub, A.J.; Chilali, M. *LMI Control Toolbox for Use with Matlab*; The MathWorks, Inc.: Massachusetts, MA, USA, 1995.

Article

Robust LQR Control for PWM Converters with Parameter-Dependent Lyapunov Functions

Nedia Aouani [1] and **Carlos Olalla** [2,*]

[1] Research Unit of System Analysis and Control, National School Engineering of Tunis, B.P. 37, Le Belvédère, Tunis 1002, Tunisia; nadia.aouani@ensta.u-carthage.tn

[2] Departament d'Enginyeria Elèctrica, Electrònica i Automàtica, Universitat Rovira i Virgili, 43007 Tarragona, Spain

* Correspondence: carlos.olalla@urv.cat; Tel.: +34-977-558-632

Received: 30 September 2020; Accepted: 23 October 2020; Published: 26 October 2020

Abstract: This paper presents a novel framework for robust linear quadratic regulator (LQR)-based control of pulse-width modulated (PWM) converters. The converter is modeled as a linear parameter-varying (LPV) system and the uncertainties, besides their rate of change, are taken into account. The proposed control synthesis method exploits the potential of linear matrix inequalities (LMIs), assuring robust stability whilst obtaining non-conservative results. The method has been validated in a PWM DC–DC boost converter, such that it has been shown, with the aid of simulations, that improved robustness and improved performance properties can be achieved, with respect to previously proposed approaches.

Keywords: uncertainty; PWM converters; LQR; LMIs; robustness; performance

1. Introduction

Control systems for power converters typically must satisfy several specifications and requirements, while dealing with uncertainty or operating point dependence at the same time. Since worst-case models may not exist or be different for each specification, the conventional industry standard approaches, such as the ones based on voltage-mode [1,2] and current-mode [2–4] controllers, rely on expert knowledge, simulation and iteration in order to find an appropriate controller.

As an alternative to this manual iteration, the automatic synthesis of controllers for switched-mode power converters has been one active topic of research in the last decade. These approaches are of interest because they can take into account the requirements together with the uncertainty or the nonlinearities of the converter to provide robust stability and performance, and they can do all that by imposing conditions beforehand.

Methods based on linear matrix inequalities (LMIs) have been some of the most successful approaches to the synthesis of robust controllers for power converters. The first attempts [5–7] demonstrated how uncertainty could be modeled and how the transient and frequency domain specifications could be taken into account. More recently, the efforts have been focused on approaches that do not require full state feedback [8], that improve the robustness [9] or the performance properties [10]. Although these papers employ averaged models of the converters, other approaches have also tackled the problem from a hybrid system perspective [11,12].

One of the open problems in the topic is the fact that the results may be conservative. The synthesized controller may not offer the best possible performance, when compared with conventionally tuned controllers, such as current-mode controllers. One possible solution to this conservativeness was shown in [13], where excellent robustness and tight regulation were achieved simultaneously, at the expense of control complexity.

One of the causes behind the conservativeness of LMI methods in [7] is the fact that the stability of the system is ensured no matter how large the derivative of the uncertain parameters may be. Specifically, when the uncertainty is characterized by being norm bounded, time varying and evolving in a set of polyhedral vertices, one difficulty remains: how to find an adequate mathematical representation for it, as well as for its rate of variation [14]. Nonetheless, several ways for representing both the derivative of the time-dependent parameter and the parameter itself have been proposed in the literature [14–17]. Different approaches to control these uncertain systems have been reported, such as state feedback gain-scheduling control [18], output feedback [15], linear quadratic Gaussian (LQG) or linear quadratic regulator (LQR) controllers [19–21] and gain-scheduled linear quadratic regulators (LQRs) [22,23].

In this paper, we propose a new method to synthesize robust LQR controllers for pulse-width modulated (PWM) converters, with the objective to improve the LQR synthesis that was proposed in [7]. The method is based on the results introduced in [14,15], such that the proposed approach can consider the time derivative of the uncertain parameters. As a consequence, the new LQR formulation can obtain less conservative results. This reduced conservativeness can be seen as a new degree of freedom. With this method, practicing engineers can synthesize controllers for larger sets of uncertainty (i.e., with improved robustness) or controllers that provide tighter regulation (i.e., improved performance) when compared with the previous method. The approach has been verified with the synthesis of a controller for a boost converter, such that a direct comparison with [7] has been carried out. Note that the proposed method could also be used in other switched-mode power converters, such as the buck converter (which was also treated in [7]).

This paper is organized as follows. Section 2 briefly reviews the modeling of the boost converter and the LQR state feedback proposed in [7]. Then, Section 3 proposes a new formulation of the LQR problem, such that novel LMI conditions are given. In Section 4, the proposed synthesis method is employed in the boost converter, using the original model and other alternatives that allow us to obtain improved robustness or improved performance. The appropriateness of the approach is verified with simulations in Section 5. Finally, conclusions are given in Section 6.

2. Modeling of the DC–DC Converter and LQR State Feedback Control

This section introduces the state feedback control approach proposed in [7], which resulted in the automatic synthesis of robust LQR controllers for PWM power converters.

2.1. Averaged Model of the DC–DC Boost Converter

Figure 1 shows the block diagram of a DC–DC converter with the control subsystem, where $v_0(t)$ is the output voltage, $v_g(t)$ is the line voltage, $i_{load}(t)$ is the load disturbance. The output voltage must be kept at a given value V_{ref}. The converter load is modeled as a resistor R.

In [7], the averaged model of a boost converter is given in the form

$$\dot{x}(t) = A(\theta)x(t) + B_u(\theta)u(t) \tag{1}$$

The uncertainty in $A(\theta)$ and $B_u(\theta)$ is included in a convex polytope as follows:

$$[A(\theta), B_u(\theta)] \quad \in Co\{\varsigma_1, \ldots, \varsigma_N\}$$
$$:= \left\{ \sum_{i=1}^{N} \lambda_i \varsigma_i \, , \, \lambda_i \geq 0, \, \sum_{i=1}^{N} \lambda_i = 1 \right\} \tag{2}$$

In general, the admissible values of vector θ are constrained in an hyperrectangle in the parameter space $\mathfrak{R}^N$.

The images of the matrix $[A(\theta), B_u(\theta)]$ for each vertex v_i correspond to a set $\{\varsigma_1, \ldots, \varsigma_N\}$. The components of the set $\{\varsigma_1, \ldots, \varsigma_N\}$ are the extrema of a convex polytope which contains the images for all admissible values of θ if the $[A(\theta), B_u(\theta)]$ depends linearly on θ [7].

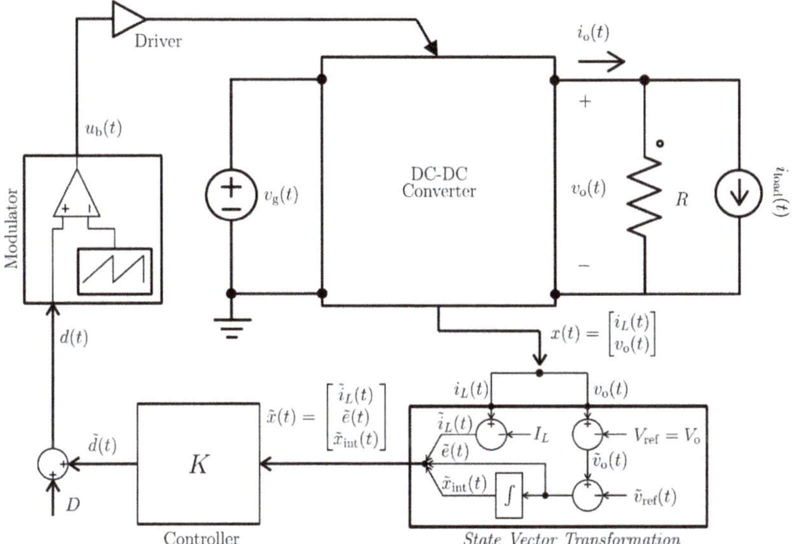

Figure 1. Schematic of a DC–DC converter with a state feedback control subsystem.

Where

$$A(\theta) = \begin{bmatrix} \frac{-R_L}{L} & -\frac{D'}{L} & 0 \\ \frac{D'}{C} & -\frac{1}{RC} & 0 \\ 0 & -1 & 0 \end{bmatrix}; \ B_u(\theta) = \begin{bmatrix} \frac{V_g}{D'L} \\ -\frac{V_g}{(D'^2 R)C} \\ 0 \end{bmatrix} \tag{3}$$

According to [7], for the DC–DC boost converter, the load R and the duty cycle D'_d at the operating point are considered uncertain parameters. Besides, two new uncertain variables, $\delta = \frac{1}{D'_d}$ and $\beta = \frac{1}{D'^2_d R}$, are defined. Thus, the parameter vector was defined as:

$$\theta = \begin{bmatrix} \frac{1}{R} & D'_d & \delta & \beta \end{bmatrix} \tag{4}$$

where the components of the parameter vector are restricted inside the following intervals:

$$\begin{aligned}
R &\in \begin{bmatrix} \frac{1}{R_{max}}, & \frac{1}{R_{min}} \end{bmatrix} \\
D'_d &\in \begin{bmatrix} D'_{dmin}, & D'_{dmax} \end{bmatrix} \\
\delta &\in \begin{bmatrix} \frac{1}{D'_{dmax}}, & \frac{1}{D'_{dmin}} \end{bmatrix} \\
\beta &\in \begin{bmatrix} \frac{1}{(D'^2_{dmax} R_{max})}, & \frac{1}{(D'^2_{dmin} R_{min})} \end{bmatrix}
\end{aligned} \tag{5}$$

This gives an uncertain model, which from now on is noted as P_{2009}, inside a polytopic domain formed by $N = 2^4$ vertices. A three-dimensional representation of P_{2009} is shown in Figure 2. This model was used in [7] to synthesize a robust LQR controller, as is explained in the next subsection.

2.2. Previous LMI Formulation of the LQR Synthesis Problem

In [7], the LQR problem was solved as follows:

$$\min_{P,Y,X} Tr(QP) + Tr(X)$$

Subject to

$$A_i P + PA_i^T + B_{ui}Y + Y^T B_{ui}^T + I < 0$$

$$\begin{bmatrix} X & R^{\frac{1}{2}}Y \\ YR^{\frac{1}{2}} & P \end{bmatrix} > 0 \tag{6}$$

$$P > 0 \, For \, i = 1, \dots, N$$

N is the number of vertices of the polytope. Q and R are constant matrices that set weights on states and control effort. Once this minimization under constraints is solved, the optimal LQR controller was recovered by $K = YP^{-1}$.

In the next section, we aim to establish new LMI formulation for an LQR problem treating linear systems with time-varying parameters.

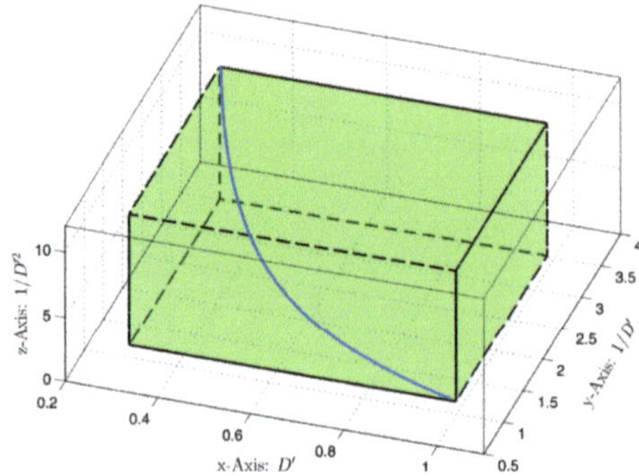

Figure 2. Plot of nonlinear uncertainty function (f(D′)) (solid line) and three-dimensional projection of the polytope P_{2009} (dashed line), as in [7].

3. New Formulation of the LQR Problem for Linear Parameter-Varying (LPV) Polytopic Systems

3.1. Proposed Representation of Uncertainty and Its Rate of Variation

Let us consider the continuous linear parameter time-dependent system, given by the state representation:

$$\dot{x}(t) = A(\theta(t))x(t) + B_u(\theta(t))u(t) \tag{7}$$

where $x(t) \in \Re^n$ is the state and $u(t) \in \Re^m$ is the input.

We assume the system matrices $A(\theta(t))$ and $B_u(\theta(t))$ are dependent on the parameter $\theta_i(t)$, i.e.,

$$A(\theta(t)) = \sum_{i=1}^{N} \theta_i(t)A_i \tag{8}$$

$$B_u(\theta(t)) = \sum_{i=1}^{N} \theta_i(t)B_{ui} \tag{9}$$

where A_i and B_i are now constant matrices ($i = 1..N$).

The time-varying parameter $\theta(t)$ varies in a polytope given by:

$$\theta(t) \in \Lambda_N, \text{ where } \Lambda_N := \left\{ \theta \in \mathfrak{R}^N : \sum_{i=1}^{N} \theta_i = 1, \ 0 \leq \theta_i \leq 1 \right\} \tag{10}$$

N is, again, the number of vertices of the polytope.

Its time derivative $\dot{\theta}(t)$ is such that:

$$\left\| \dot{\theta}(t) \right\| \leq b; \ \ b \geq 0 \tag{11}$$

b is a positive real number that bounds the parameter's derivative.

If the uncertain parameter θ_i belongs to the set given by (10) and satisfies (11), then its time derivative can be written as [14,15]:

$$\dot{\theta}_i = r(\sigma_j - \beta_k) \tag{12}$$

σ_j and β_k belong, respectively, to the polytopes given by:

$$\sigma(t) \in \Lambda_M; \ \ \Lambda_M := \left\{ \sigma \in \mathfrak{R}^M : \sum_{j=1}^{M} \sigma_j = 1, \ 0 \leq \sigma_j \leq 1 \right\} \tag{13}$$

$$\beta(t) \in \Lambda_K; \ \ \Lambda_K := \left\{ \beta \in \mathfrak{R}^K : \sum_{k=1}^{K} \beta_k = 1, \ 0 \leq \beta_k \leq 1 \right\} \tag{14}$$

3.2. New LQR Problem Formulation for Uncertain LPV System

We are interested in an LMI formulation of the LQR problem adapted from [7]. Given the system presented in (1), the optimal LQR controller is obtained by using the state feedback gain K ($u = Kx$) that minimizes a performance index.

$$J = \int_0^{\infty} \left(x^T Q x + u^T R u \right) dt \tag{15}$$

where Q is a symmetric and semidefinite positive matrix and R is a symmetric and definite positive matrix.

The pair (A, Bu) must be controllable. The LQR problem can be viewed as the weighted minimization of a linear combination of the state x and the control input u. The weighting matrix Q establishes which states are to be controlled more tightly than others. R weights the amount of control action to be applied depending on how large the deviation of the state x is [7]. This optimization of cost weight constrains the magnitude of the control signal. The LQR controller is obtained by using the feedback gain K such that, in closed loop, the performance index (15) is rewritten:

$$J = \int_0^{\infty} \left(x^T (Q + K^T R K) x \right) dt \tag{16}$$

In this paper, we aim to give an LMI formulation for the same LQR problem as in [7], taking into account the uncertain parameter $\theta(t)$ that evolves into (10). We also consider the time derivative of this parameter as it is expressed in (12). The novel LQR formulation for the LPV system is given in the following theorem.

Theorem 1. *The complete LMI formulation of the LQR problem is: considering system (7), in the uncertain domains (10)–(11), with Nsymmetric and positive definite matrices $P_1, \ldots .P_N$, N matrices $F_j, G_j (j = 1, \ldots N)$, matrices L and R of appropriate dimensions and a positive realα that is sufficiently large, we have:*

$$\min_{P_i, F_j, G_j, L, R} \left[\sum_{i=1}^{i=N} Tr(QP_i) + R(\delta + \sigma) \right] \tag{17}$$

Subject to

$$\begin{pmatrix} b(P_j - P_k) - \alpha F_j - \alpha F_j^T + I & -\alpha G_j - F_j + P_i & L^T A_i^T + D^T B_{ui}^T + \alpha L^T + F_j \\ -\alpha G_j - F_j^T + P_i^T & -G_j - G_j^T & G_j \\ A_i L + B_{ui} D + \alpha L + F_j^T & G_j^T & -L - L^T \end{pmatrix} < 0 \tag{18}$$

$$\begin{bmatrix} \sigma I_{n\times n} & D^T \\ D & I_{m\times m} \end{bmatrix} > 0$$

$$\begin{bmatrix} \delta I_{n\times n} & I_{n\times n} \\ I_{n\times n} & L \end{bmatrix} > 0 \tag{19}$$

$$P_i > 0 \tag{20}$$
$$For\ i, j, k = 1, .., N$$

The control gain is then given by $K = DL^{-1}$.

Proof. Let us consider (18); replacing D by KL and D^T by $L^T K^T$ in (18), we get:

$$\begin{pmatrix} b(P_j - P_k) - \alpha F_j - \alpha F_j^T + I & -\alpha G_j - F_j + P_i & L^T A_i^T + L^T K^T B_{ui}^T + \alpha L^T + F_j \\ -\alpha G_j - F_j^T + P_i^T & -G_j - G_j^T & G_j \\ A_i L + B_{ui} KL + \alpha L + F_j^T & G_j^T & -L - L^T \end{pmatrix} < 0 \tag{21}$$

In (21), replacing $A_i + B_{ui} K$ by A_i and $A_i^T + K^T B_{ui}^T$ by A_i^T, we get:

$$\begin{pmatrix} b(P_j - P_k) - \alpha F_j - \alpha F_j^T + I & -\alpha G_j - F_j + P_i & L^T A_i^T + \alpha L^T + F_j \\ -\alpha G_j - F_j^T + P_i^T & -G_j - G_j^T & G_j \\ A_i L + \alpha L + F_j^T & G_j^T & -L - L^T \end{pmatrix} < 0 \tag{22}$$

Multiplying (22) by θ_i, σ_j and β_k and summing up, respectively, for $i = 1 \ldots N$, $j = 1..M$ and $k = 1..K$, we obtain:

$$\begin{pmatrix} b(P(\sigma) - P(\beta)) - \alpha F(\sigma) - \alpha F^T(\sigma) + I & -\alpha G(\sigma) - F(\sigma) + P(\theta) & \begin{matrix} L^T A^T(\theta) + \alpha L^T \\ + F(\sigma) \end{matrix} \\ -\alpha G(\sigma) - F^T(\sigma) + P^T(\theta) & -G(\sigma) - G^T(\sigma) & G(\sigma) \\ \begin{matrix} A(\theta)L + \alpha L \\ + F^T(\sigma) \end{matrix} & G^T(\sigma) & -L - L^T \end{pmatrix} < 0 \tag{23}$$

Multiplying the LMI condition (23) by $\begin{pmatrix} I & 0 & \alpha I \\ 0 & I & 0 \\ 0 & 0 & I \end{pmatrix} < 0$ on the left and its transpose on the right,

where α is a positive real number, we get:

$$\begin{pmatrix} \dot{P}(\theta) + \alpha A(\theta)L + \alpha L^T A^T(\theta) + I & -F(\sigma) + P(\theta) & L^T A^T(\theta) - \alpha L + F(\sigma) \\ -F^T(\sigma) + P^T(\theta) & -G(\sigma) - G^T(\sigma) & G(\sigma) \\ A(\theta)L - \alpha L^T + F^T(\sigma) & G^T(\sigma) & -L - L^T \end{pmatrix} < 0 \tag{24}$$

where $P(\theta)$ is a positive and symmetric matrix called the Lyapunov candidate matrix.

We suppose that $\alpha L^T = P(\theta) = F(\sigma)$ and $G(\sigma) = \frac{P(\theta)}{\alpha}$, then, we get:

$$\begin{pmatrix} \dot{P}(\theta) + A(\theta)P(\theta) + P(\theta)A^T(\theta) + I & 0 & \frac{P(\theta)A^T(\theta)}{\alpha} \\ 0 & -\frac{2P(\theta)}{\alpha} & \frac{P(\theta)}{\alpha} \\ \frac{A(\theta)P(\theta)}{\alpha} & \frac{P(\theta)}{\alpha} & -L - L^T \end{pmatrix} < 0 \tag{25}$$

Applying the Schur complement on LMI (25),

$$\dot{P}(\theta) + A(\theta)P(\theta) + P(\theta)A^T(\theta) + I < -\frac{2}{4\alpha - 1}A^T(\theta)P(\theta)A(\theta) \tag{26}$$

For values of α that are sufficiently large, we get

$$\dot{P}(\theta) + A(\theta)P(\theta) + P(\theta)A^T(\theta) + I < 0 \tag{27}$$

(27) can be written

$$\dot{P}(\theta) + A(\theta)P(\theta) + P(\theta)A^T(\theta) < -I \tag{28}$$

Thus, we get the Lyapunov condition written for the LPV systems

$$\dot{P}(\theta) + A(\theta)P(\theta) + P(\theta)A^T(\theta) < 0 \tag{29}$$

For the proof of (19), see [24].

The approach presented above is used for the case of a boost DC–DC converter modeled based on an LPV polytopic formulation.

4. Synthesis of Improved LQR Controllers for DC–DC Boost Converters

4.1. Modeling

In this section, two different uncertainty models are shown. The same uncertain parameter $\theta(t)$ is employed. The uncertain parameter belongs to (4) and is such that its derivative verifies (5) and (6).

$$\theta(t) = \left[D', \frac{1}{D'}, \frac{1}{R}, \frac{1}{D'^2 R} \right] \tag{30}$$

Any matrix in this set can be obtained by:

$$A(\theta(t)), B_u(\theta(t)) = \theta_1(A_1, B_{u1}) + \theta_2(A_2, B_{u2}) + \theta_3(A_3, B_{u3}) + \theta_4(A_4, B_{u4}) \tag{31}$$

and the derivative of $\theta(t)$ satisfies the bound imposed in Section 2.

The first model is a simplification of P_{2009}, and it was first introduced in [25]. This model, which will be noted as P_{2011}, is based on a polytopic covering of the space in $\theta(t)$. Since the variables in $\theta(t)$ are not fully independent, a polytopic covering with fewer vertices can be derived. The result is a polytope with eight vertices instead of the 16 vertices in P_{2009}. Figure 3 shows P_{2011} and Table 1 defines its vertices.

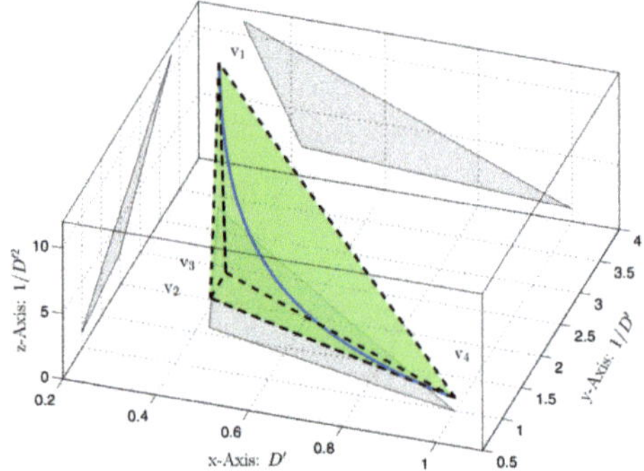

Figure 3. Plot of nonlinear uncertainty function (f(D′)) (solid line) and reduced polytope P_{2011} (dashed line), as in [25]. The projections of the polytope are also shown in each respective plane.

Table 1. Vertices of the polytopic covering of P_{2011}

	D'	$\frac{1}{D'}$	$\frac{1}{D'^2R}$	$\frac{1}{R}$
$\theta_1(t)$	0.3	1/0.3	1/0.9	1/10
$\theta_2(t)$	0.3	1/0.3	1/4.5	1/50
$\theta_3(t)$	0.425	1.6	2.25/10	1/10
$\theta_4(t)$	0.425	1.6	2.25/50	1/50
$\theta_5(t)$	0.425	2	2.25/10	1/10
$\theta_6(t)$	0.425	2	2.25/50	1/50
$\theta_7(t)$	1	1	1/10	1/10
$\theta_8(t)$	1	1	1/50	1/50

In order to test if the proposed synthesis approach can extend the region of stability of the system, we consider an extension of P_{2011}. This is a new model that considers an enlargement of the space in $\theta(t)$. Figure 4 shows the original P_{2011} polytope, and the novel enlarged one, noted as P_{2020}. The vertices of the model are shown in Table 2.

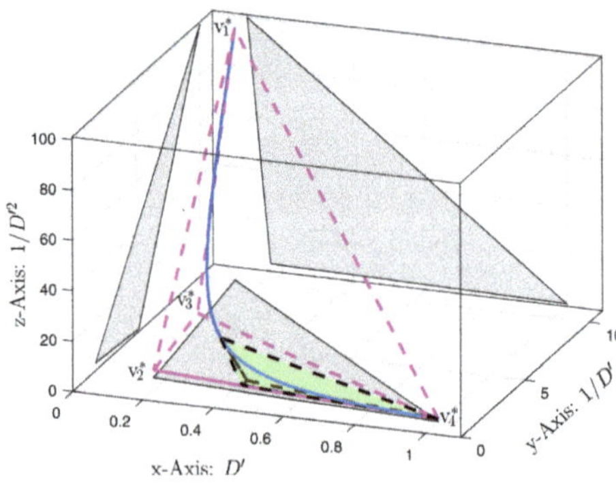

Figure 4. Plot of nonlinear uncertainty function (f(D′)) (solid line) and proposed polytope P_{2020} for extended robustness (dashed pink line). The projections of the polytope are also shown in each respective plane. Note how P_{2020} compares with P_{2011}, which is shown here in black dashed lines and covers a much smaller parameter space.

Table 2. Vertices of the polytopic covering of P_{2020}

	D'	$\frac{1}{D'}$	$\frac{1}{D'^2 R}$	$\frac{1}{R}$
$\theta_1(t)$	0.1	1/0.1	1/0.1	1/10
$\theta_2(t)$	0.1	1/0.1	1/0.5	1/50
$\theta_3(t)$	0.165	1.8	3/10	1/10
$\theta_4(t)$	0.165	1.8	3/50	1/50
$\theta_5(t)$	0.165	5.2	1	1/10
$\theta_6(t)$	0.165	5.2	1/5	1/50
$\theta_7(t)$	1	1	1/10	1/10
$\theta_8(t)$	1	1	1/50	1/50

4.2. Synthesis Results

The same Q used in [7] is employed in the synthesis. The value of R was established using the proposed synthesis method with the old model P_{2009}. The objective was to obtain a controller that is equivalent to the one in [7], which is noted K_{eq}. That aim was achieved with $R = 1 \cdot 10^{-6}$.

4.2.1. Previous LQR Synthesis Method

Based on the LQR synthesis method given in [7], whose LMIs are shown in (6), three models have been tested: P_{2009}, P_{2011} and P_{2020}. The results are as follows:

- With P_{2009}:$K = \begin{bmatrix} -0.86 & -1.39 & 3159.54 \end{bmatrix}$.
- With P_{2011}: The set of LMIs is infeasible.
- With P_{2020}: The set of LMIs is infeasible.

4.2.2. Proposed LQR Synthesis Method

The synthesis with the novel conditions {17-18-19-20} results in the following controllers:

- With P_{2009}, $b = 1 \cdot 10^4$, $\alpha = 1 \cdot 10^5$, $R = 1 \cdot 10^{-6}$, the result is $K_{eq} = \begin{bmatrix} -0.55 & -0.89 & 1871.75 \end{bmatrix}$.

This controller achieves the same performance that can be obtained with controller K, but with a lower control effort (the gains in K_{eq} are smaller than those in K).

- With P_{2011}, $b = 1 \cdot 10^4$, $\alpha = 1 \cdot 10^5$, $R = 1 \cdot 10^{-6}$, the result is $K_{perf} = \begin{bmatrix} -0.46 & -1.49 & 4218 \end{bmatrix}$.
- With P_{2020}, $b = 1 \cdot 10^4$, $\alpha = 1 \cdot 10^5$, $R = 1 \cdot 10^{-5}$, the result is $K_{rob} = \begin{bmatrix} -0.01262 & 0.00095 & 9.607 \end{bmatrix}$.

5. Simulation Results

This section illustrates the properties of the different controllers K, K_{eq}, K_{perf} and K_{rob}. We have performed a set of PSIM [26] simulations of the switched DC–DC boost converter, according to Figure 1. The first set of simulations is useful to establish the performance of the controllers, by analyzing the response of the converter with respect to changes in the output current. The second set aims to establish the robustness of the different controllers when there is a change in the operating point, by modifying the supply voltage.

First, the waveforms of the simulations with changes in the load are grouped in Figure 5. The top waveforms in each subfigure correspond to the output voltage $v_0(t)$, whereas the bottom waveform represents the output current $i_{load}(t)$. In all simulations, the converter load is initially the nominal value $R = 25\Omega$. At time $t = 1$ ms, the load changes to $R = 10\Omega$, which is the maximum load allowed by design in all polytopes. The load returns to $R = 25\Omega$ at $t = 6$ ms.

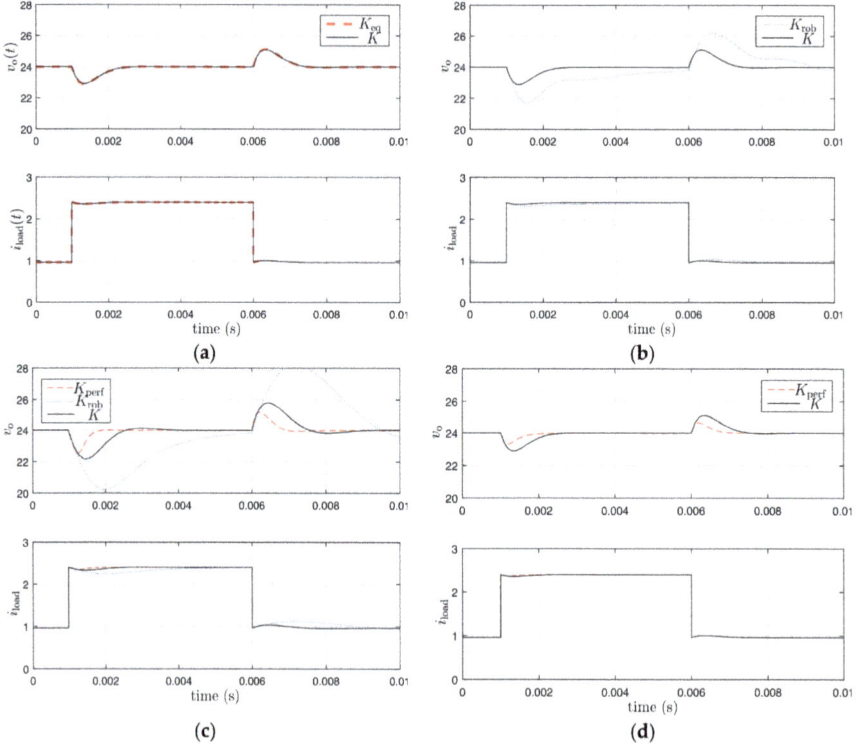

Figure 5. Simulated transient of the boost converter under a load step transient with the robust LQR controllers K (solid line), K_{eq} (thick dashed line), K_{perf} (dashed line) and K_{rob} (dotted line); (**a–c**) show the transient when Vg = 12 V and the operating point duty cycle is D′ = 0.5; (**d**) shows the transient when Vg = 7.2 V and the operating point duty cycle is D′ = 0.3.

As a baseline for the comparison, Figure 5a shows the performance of controller K, as in [7], and the performance of controller K_{eq} obtained with the proposed method and the same polytope used in [7], P_{2009}. It can be seen that the disturbance rejection properties and the settling time are nearly identical. In contrast, Figure 5b shows a comparison with controller K_{perf}, which exhibits a tight regulation of the output voltage, such that the maximum error of $v_0(t)$ and its settling time are reduced to approximately one half of what is achieved with K. As expected, the robust controller K_{rob} presents loose regulation and a slower response, as shown in Figure 5c, when compared to K.

Note that Figure 5a–c shows the response at the nominal operating point, when $v_g(t) = 12V$ and D′ = 0.5. In order to evaluate the performance at a different operating point, Figure 5d shows the response of K, K_{perf} and K_{rob} under an input voltage variation of −40%, such that the operating point is now D′ = 0.3. Again, K_{perf} is the controller that achieves excellent regulation properties, maintaining its robustness in the expected region of operation.

If K_{perf} is the controller that demonstrates that the proposed method can be used to improved regulation while maintaining the same robustness properties, K_{rob} is the controller that demonstrates that the method can also be employed to enlarge the stability region. Figure 6 shows the waveforms of the simulations in which the input voltage is stepped, such that the operating point of the converter is modified in time. Figure 6a shows a voltage step of −40%, which corresponds to a step in the duty cycle from D′ = 0.5 to D′ = 0.3 (D = 0.7). Since all polytopes considered such a region, the three controllers maintain the stability, with K_{perf} exhibiting the best regulation performance. Figure 6b shows a similar step in the input voltage, but now the input voltage decreases down to $v_g(t) = 2.4V$, such that the operating point duty cycle moves from D' = 0.3 to D' = 0.1. The method proposed in [7] did not allow us to consider such a large range of operating point uncertainty, whereas the proposed method resulted in controller K_{rob}. As can be seen in the figure, K_{rob} is the only controller that successfully maintains stability under those conditions, exhibiting excellent stability properties.

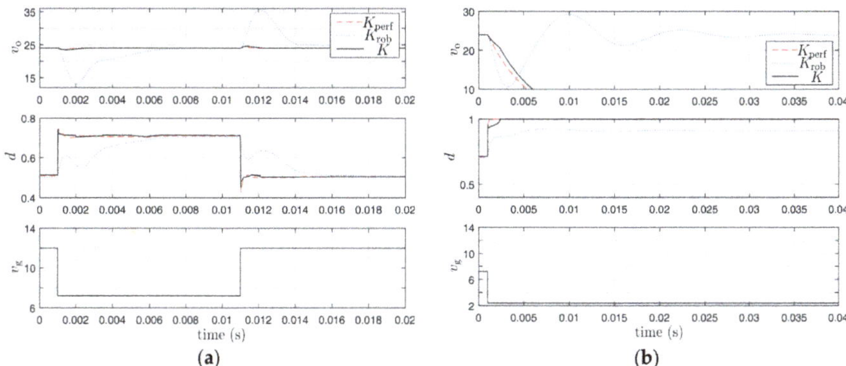

Figure 6. Simulated transient of the boost converter, in the presence of an input voltage disturbance, with the robust LQR controllers K (solid line), K_{perf} (dashed line) and K_{rob} (dotted line). (**a**) The input voltage steps down to 7.2 V, which corresponds to to D = 0.7. All controllers consider such a change of operating point and maintain the stability. (**b**) The input voltage steps down to 2.4 V, which corresponds to D = 0.9. Only controller K_{rob} maintains the stability of the regulation.

It is worth noting that the transient shown in Figure 6b shows the saturation of the duty cycle at 100% with the unstable controllers. Although the modeling of that nonlinearity is out of the scope of this paper, this aspect has been treated in the specific context of switched-mode power converters in [27].

Finally, Figure 7 depicts the waveforms of the converter startup, with the three controllers K, K_{perf} and K_{rob}. The input voltage is Vg = 12 V and the voltage reference ramps up from 12 V to 24 V at $t = 0$, with a rate of change of 2400 V/s. It can be observed that the three controllers operate inside the expected range of operation and stabilize the converter.

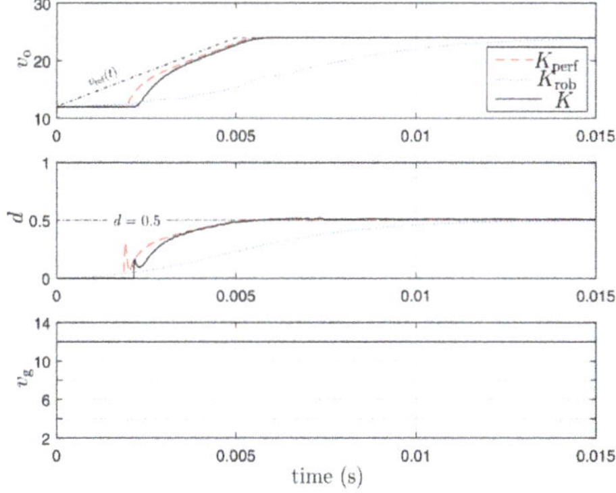

Figure 7. Simulated transient of the boost converter during startup with Vg = 12 V, for the three controllers K (solid line), K_{perf} (dashed line) and K_{rob} (dotted line). Top waveforms: output voltage. Middle waveforms: duty cycle. Bottom waveform: input voltage.

6. Conclusions

The numerical synthesis of robust LQR controllers for PWM DC–DC converters by means of LMIs has suffered from the conservativeness of the methods based on quadratic stability, since a single Lyapunov function is employed for the entire uncertainty region and because the uncertain parameters are assumed to change arbitrarily fast. This paper proposes a new method to synthesize robust LQR controllers. The method employs parameter-dependent Lyapunov functions and allows us to consider the rate of change of the uncertain parameters.

The method has been employed to synthesize LQR controllers for a PWM DC–DC boost converter. With that aim, the paper has reviewed two uncertainty models of the boost converter that were proposed in the past. In addition, it has introduced an enlarged version of one of them, with the objective to obtain stability for a very large region of uncertain parameters. While the conventional synthesis methods fail to obtain feasible solutions with these uncertainty models, the proposed method has been demonstrated to be useful in achieving better regulation performance or improved robustness.

Author Contributions: N.A. and C.O. conceptualized the research; N.A and C.O. conceived the methodology and the formal analysis; C.O. completed the investigation and carried out validation tasks; N.A. and C.O. wrote the original draft, reviewed and edited the final paper. All authors have read and agreed to the published version of the manuscript.

Funding: The research leading to these results has received funding from the Spanish Ministry of Economy and Competitiveness under grant DPI2017-84572-C2-1-R, and also from the Research Unit of System Analysis and Control, National School Engineering of Tunis, Tunisian Ministry of Scientific Research and Higher Education.

Conflicts of Interest: The authors declare no conflict of interest.

References

1. Middlebrook, R.D.; Ćuk, S. A general unified approach to modelling switching-converter power stages. *Int. J. Electron.* **1977**, *42*, 521–550. [CrossRef]
2. Erickson, R.W.; Maksimović, D. *Fundamentals of Power Electronics*; Springer: Cham, Switzerland, 2001.
3. Deisch, C. Simple switching control method changes power converter into a current source. *IEEE Power Electron. Spec. Conf.* **1978**, 300–306. [CrossRef]
4. Capel, A.; Ferrante, G.; O'Sullivan, D.; Weinberg, A. Application of the injected current model for the dynamic analysis of switching regulators with the new concept of LC3 modulator. *IEEE Power Electron. Spec. Conf.* **1978**, 135–147. [CrossRef]
5. Montagner, V.; Peres, P. Robust pole location for a DC-DC converter through parameter dependent Control In Proceedings of the 2003 International Symposium on Circuits and Systems, ISCAS 2003. Bangkok, Thailand, 25–28 May 2003; Volume 3, pp. 351–354.
6. Olalla, C.; Leyva, R.; el Aroudi, A.; Garces, P.; Queinnec, I. LMI robust control design for boost PWM converters. *IET Power Electron.* **2010**, *3*, 75. [CrossRef]
7. Olalla, C.; Leyva, R.; el Aroudi, A.; Queinnec, I. Robust LQR Control for PWM Converters: An LMI Approach. *IEEE Trans. Ind. Electron.* **2009**, *56*, 2548–2558. [CrossRef]
8. Gabe, I.J.; Montagner, V.F.; Pinheiro, H. Design and Implementation of a Robust Current Controller for VSI Connected to the Grid Through an LCL Filter. *IEEE Trans. Power Electron.* **2009**, *24*, 1444–1452. [CrossRef]
9. Olalla, C.; Queinnec, I.; Leyva, R.; el Aroudi, A. Optimal State-Feedback Control of Bilinear DC–DC Converters With Guaranteed Regions of Stability. *IEEE Trans. Ind. Electron.* **2011**, *59*, 3868–3880. [CrossRef]
10. Maccari, L.A.; Massing, J.R.; Schuch, L.; Rech, C.; Pinheiro, H.; Oliveira, R.C.L.F.; Montagner, V.F. LMI-Based Control for Grid-Connected Converters With LCL Filters Under Uncertain Parameters. *IEEE Trans. Power Electron.* **2013**, *29*, 3776–3785. [CrossRef]
11. Albea-Sanchez, C.; Garcia, G. Robust hybrid control law for a boost inverter. *Control Eng. Pr.* **2020**, *101*, 104492. [CrossRef]
12. Sferlazza, A.; Albea-Sanchez, C.; Garcia, G. A hybrid control strategy for quadratic boost converters with inductor currents estimation. *Control Eng. Pr.* **2020**, *103*, 104602. [CrossRef]
13. Olalla, C.; Leyva, R.; Queinnec, I.; Maksimovic, D. Robust Gain-Scheduled Control of Switched-Mode DC–DC Converters. *IEEE Trans. Power Electron.* **2012**, *27*, 3006–3019. [CrossRef]

14. Aouani, N.; Salhi, S.; Ksouri, M.; Garcia, G. H2 analysis for LPV systems by parameter-dependent Lyapunov functions. *IMA J. Math. Control Inf.* **2011**, *29*, 63–78. [CrossRef]
15. Aouani, N.; Salhi, S.; Garcia, G.; Ksouri, M. New robust stability and stabilizability conditions for linear parameter time varying polytopic systems. In Proceedings of the 2009 3rd International Conference on Signals, Circuits and Systems (SCS), Medenine, Tunisia, 6–8 November 2009; pp. 1–6.
16. Xie, W. H2 gain scheduled state feedback for LPV system with new LMI formulation. *IEEE Proc. Control Theory Appl.* **2005**, *152*, 693–697. [CrossRef]
17. Cao, Y.-Y.; Lin, Z. A Descriptor System Approach to Robust Stability Analysis and Controller Synthesis. *IEEE Trans. Autom. Control* **2004**, *49*, 2081–2084. [CrossRef]
18. Wu, F.; Grigoriadis, K.M. LPV Systems with parameter-varying time delays: Analysis and Control. *Automatica* **2001**, *37*, 221–229. [CrossRef]
19. Wu, F.; Packard, A. LQG control design for LPV systems. In Proceedings of the 1995 American Control Conference—ACC'95, Seattle, WA, USA, 21–23 June 1995.
20. Rödönyi, G.; Bokor, J. LQG control of LPV systems with parameter-dependent Lyapunov function. In Proceedings of the 10th Mediterranean Conference, on Control and Automation—MED2002, Lisbon, Portugal, 9–12 July 2002.
21. Osório, C.; Koch, G.; Oliveira, R.C.; Montagner, V.F. A practical design procedure for robust H2 controllers applied to grid-connected inverters. *Control Eng. Pr.* **2019**, *92*. [CrossRef]
22. Liu, Z.; Theilliol, F.; Gu, D.; He, Y.; Yang, L.; Han, J. State Feedback Controller Design for Affine Parameter-Dependent LPV systems. *IFAC PapersOnLine* **2017**, *50*, 9760–9765. [CrossRef]
23. Ilka, A.; Vesely, V. Robust LPV-based infinite horizon LQR design. In Proceedings of the 2017 21st International Conference on Process Control (PC), Štrbské Pleso, Slovakia, 6–9 June 2017; pp. 86–91.
24. Tarbouriech, S.; Souères, P.; Gao, B. A multicriteria image-based controller based on a mixed polytopic and norm-bounded representation of uncertainties. In Proceedings of the 44th IEEE Conference on Decision and Control, Seville, Spain, 12–15 December 2005; pp. 5385–5390.
25. Olalla, C.; Queinnec, I.; Leyva, R.; el Aroudi, A. Robust optimal control of bilinear DC–DC converters. *Control Eng. Pr.* **2011**, *19*, 688–699. [CrossRef]
26. PSIM User Manual, Powersim Inc. 2003. Available online: https://powersimtech.com/drive/uploads/2019/04/PSIM-user-manual.pdf (accessed on 30 August 2020).
27. Olalla, C.; Leyva, R.; el Aroudi, A.; Queinnec, I.; Tarbouriech, S. Hinf Control of DC-DC Converters with Saturated Inputs. In Proceedings of the IEEE Annual Conference on Industrial Electronics IECON, Porto, Portugal, 3–5 November 2009.

Publisher's Note: MDPI stays neutral with regard to jurisdictional claims in published maps and institutional affiliations.

applied
sciences

Article

ADC Quantization Effects in Two-Loop Digital Current Controlled DC-DC Power Converters: Analysis and Design Guidelines

Catalina González-Castaño [1], Carlos Restrepo [2], Roberto Giral [1], Enric Vidal-Idiarte [1,*] and Javier Calvente [1]

1   Departament d'Enginyeria Electrònica, Elèctrica i Automàtica, Escola Tècnica Superior d'Enginyeria, Universitat Rovira i Virgili, 43007 Tarragona, Spain; gocafe1@gmail.com (C.G.-C.); roberto.giral@urv.cat (R.G.); javier.calvente@urv.cat (J.C.)
2   Department of Electromechanics and Energy Conversion, Universidad de Talca, Curicó 3340000, Chile; crestrepo@utalca.cl
*   Correspondence: enric.vidal@urv.cat; Tel.: +34-977-559622

Received: 23 September 2020; Accepted: 9 October 2020; Published: 15 October 2020

Abstract: This paper analyzes the presence of undesired quantization-induced perturbations (QIP) in a dc-dc buck-boost converter using a two-loop digital current control. This work introduces design conditions regarding control laws gains and signal quantization to avoid the quantization effects due to the addition of the outer voltage loop in a digital current controlled converter. The two-loop controller is composed of a multisampled average current control (MACC) in the inner current-programmed loop and a proportional-integrator compensator at the external loop. QIP conditions have been evaluated through simulations and experiments using a digitally controlled pulse width modulation (DPWM) buck-boost converter. A 400 V 1.6 kW proof-of-concept converter has been used to illustrate the presence of QIP and verify the design conditions. The controller is programmed in a digital signal controller (DSC) TMS320F28377S with a DPWM with 8.96-bit equivalent resolution, a 12-bit ADC for current sampling, and a 12-bit ADC for voltage sampling or a 16-bit ADC for voltage error sampling.

Keywords: dc-dc power converter; multisampled average current control (MACC); digital control; limit-cycle oscillation (LCO); quantization-induced perturbations (QIP)

---

1. Introduction

Digital control in dc-dc converters is of interest because its many potential advantages such as low power consumption and flexibility to program and design advanced control strategies to improve the system performance [1–3]. Therefore, the digital closed-loop configuration is increasingly being used in dc-dc converters [4–6]. Digital control depicts an important element of power converters for renewable energy systems [7], automobile industry [8], and industrial applications [9]. However, many works report disadvantageous quantization effects related to the existence of limit cycles in digitally controlled pulse width modulation (DPWM) converters. Static and dynamic models taking into account the quantization effects are derived and used to explain the origins of limit-cycle oscillations (LCO) for voltage single-loop digital control in [10,11]. A DPWM resolution lower than ADC resolution usually causes LCO that affects the regulation of the controlled variable [12]. Therefore, DPWM with resolution higher than the ADC is usually implemented in order to reduce the effect of limit-cycle oscillations in voltage single-loop digital

Appl. Sci. **2020**, *10*, 7179; doi:10.3390/app10207179

control, where the difference between the voltage reference and the output voltage is quantized to a digital number to represent the error signal [13,14].

It is well known that cascade control of dc-dc converters generally offers better performance than single-loop control [15,16]. Moreover, current control strategies are always needed to connect in parallel some converters to increase power management. Therefore, two-loop digital control structures have been extensively applied last teen years [17–21]. Analysis of LCO are given in [22–24] for digital current mode control. In these works, the resolution of the DPWM is also greater than the ADC resolution in the outer voltage loop. In [22], an estimation algorithm has been applied to the average current control of a buck converter in order to reduce quantization effects in the inductor current loop and, consequently, the presence of limit cycle oscillations. A method to design a two-loop digital control is developed in [23], where the current reference is dynamically adjusted to give a solution to the LCO problem. A technique to compute the steady-state duty cycle in real-time was considered in [24], where a time-to-digital converter translates the duty ratio information into a digital code using a moving average filter and an adjustable current loop sampling frequency. At steady-state, the strategy disables the current-loop sampling and the control computation. Then, a virtual open loop configuration is used to reduce oscillations of the inductor current.

In order to improve the resolution of the DPWM in single-loop digital voltage controllers, some authors use sigma-delta modulation to eliminate the quantization noise and the LCO. In [12], a non-zero error method is used to encode the output voltage error improving the low resolution of the DPWM. A sigma delta modulation scheme and switching frequency modulation strategy are combined in [25] to increase the effective resolution of the DPWM. Nonetheless, there are not reported works that show the effects of quantization in dc-dc converter with a two-loop digital control having an integral term of its output voltage error.

This paper presents design conditions to avoid the effects of the quantization in two-loop current controlled dc-dc switching converter. LCOs conditions presented in [10] are extended to a two-loop digital control in order to obtain restrictions associated with the gains of the control laws and the quantization resolution for each control loop. When the condition proposed for the external loop is fulfilled, simulation and experimental results verify that the QIP are suppressed from the current signals.

Section 2 presents the conditions for each digital loop to observe the two-loop quantization effects. Section 3 describes the implementation of the digitally controlled buck-boost converter in order to validate the restrictions. Experimental and simulation results for a 400 V 1.6 kW digitally controlled coupled-inductor dc-dc buck-boost converter are presented in Section 4. Conclusions are given in Section 5.

2. Two-Loop Quantization Effects

Quantization effects have been deeply studied in [10,11,26], where authors studied limit cycling conditions regarding plant and controller gains besides ADC and PWM resolution in a single-loop voltage control. This section presents dynamic conditions to avoid limit cycles in a two-loop digital current controller converter.

In this case, both inner and outer small-signal representation of the control to output transfer function can be represented in general form as

$$G_p(s) = \frac{G}{s} \tag{1}$$

where G is the gain of the transfer function and $1/s$ represents the transfer function of an integrator. In the case of the inner current control (see Figure 1a), the transfer Function (1) becomes

$$G_{pin}(s) = \frac{m_1 + m_2}{s} \tag{2}$$

where m_1 is the positive slope and m_2 is the negative slope of the output current. Finally, for the voltage loop (see Figure 1b) the transfer Function (1) becomes

$$G_{pou}(s) = \frac{1}{C_o s} \tag{3}$$

where C_o represents the output filter capacitor of the converter. A standard Proportional-Integral (PI) controller is used in both control loops

$$G_{pi}(s) = K_p + \frac{K_i}{s} \tag{4}$$

where $G_{pi}(s) = G_{pii}(s)$ for the inner loop (Figure 1a) and $G_{pi}(s) = G_{piv}(s)$ for the external loop (Figure 1b).

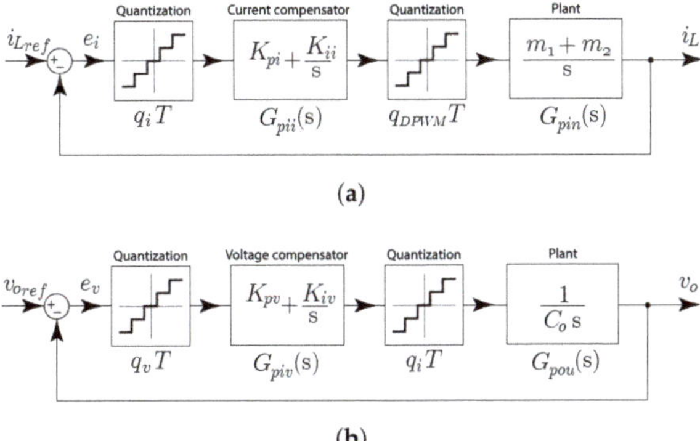

(a)

(b)

Figure 1. Proportional-integrator control block diagram of (**a**) inner loop dynamic model and (**b**) outer loop dynamic model.

The dynamic model for each loop is shown in Figure 1, where q is the quantization level for an input or an output signal. Therefore, q_i, q_v, and q_{DPWM} represent, respectively, the output current i_L, the output voltage v_o, and the DPWM quantization level. T is the switching period $(1/f_s)$. Finally, e_v and e_i are the error signals of the measured voltage and current, respectively. Then, in the inner loop, $G_{pii}(s)$ generates the control variable u, taking into account the mean value of the output current converter to change the duty cycle. Nonetheless, $G_{piv}(s)$ for the output voltage gives the current reference for the inner loop based on the error voltage.

The loop gains of the linear part of the system are defined without quantization [10,26] as follows,

$$T_L(s) = G_{pi}(s)G_p(s). \tag{5}$$

Therefore, using (1) and (4) we obtain the crossover angular frequency as

$$\omega_c = K_p G, \tag{6}$$

A typical design of PI parameters usually places the zero of the controller at least one decade below of the desired crossing frequency ω_c, thus giving the following condition,

$$K_i / K_p \ll \omega_c, \tag{7}$$

Then, we have to adjust K_p and K_i in order to obtain the desired phase margin. Phase margin (PM) is usually adjusted to be greater than $50°$ by tuning K_p and satisfying (7).

2.1. Outer Loop Condition

In digital power converter operation, the static and the integral gain condition must be satisfied to avoid limit cycling due to quantization [27–29]. Then, the necessary no-limit-cycling condition that allows the existence of a steady-state solution inside ADC zero-error bin, given in [28,29], can be written for the external voltage loop in Figure 1b as

$$q_i TG < q_v. \tag{8}$$

Condition (8) indicates that the minimum output voltage variation, due to the minimum output current step change provoked by a variation in the output voltage, must be smaller than the quantization level of the output voltage. In this condition (8), the gain G is defined as $1/C_o$. The no-limit-cycling condition involving the integral gain is

$$q_v TK_{iv} < q_i. \tag{9}$$

The output current reference change provoked by a minimum error in the voltage loop is $K_{pv}q_v$. To guarantee output voltage regulation, the compensator must develop a correction action when e_v is different from 0 [29], thus giving

$$q_v K_{pv} > q_i. \tag{10}$$

Combining restrictions (9) and (10) results in the following condition,

$$K_{iv}T < \frac{q_i}{q_v} < K_{pv}. \tag{11}$$

Condition (12) is derived replacing (10) in (8)

$$K_{pv}TG < 1. \tag{12}$$

Employing Equation (6) and replacing $K_p = K_{pv}$ in (12), we obtain an upper limit for the crossing frequency

$$\omega_c < \frac{1}{T}. \tag{13}$$

2.2. Inner Loop Condition

Restriction (11) can be extended in terms of the inner loop block diagram representation of Figure 1a. Following the same procedure as in the outer loop, the condition for the inner loop is given by

$$K_{ii}T < \frac{q_{DPWM}}{q_i} < K_{pi}, \tag{14}$$

Following (7), we select $K_{ii}T = K_{pi}/10$ and adjust K_{pi} to obtain a PM greater than 50°.

3. Validation of the Restrictions

The fulfillment of Conditions (11) and (14), which guarantee a stable digital two-loop control, have been verified using a buck-boost converter with coupled inductors. The topology of the dc-dc buck-boost converter for a voltage regulation application shown in Figure 2 was introduced as an unidirectional buck-boost converter in [30] and presented for electric vehicle and high-voltage application in [31,32]. The bidirectional power stage shown in Figure 2 is composed of two coupled inductors with unitary turns ratio and magnetic coupling coefficient $k = 0.5$. Therefore, primary self-inductance L_1 is equal to secondary self-inductance L_2 ($L_1 = L_2 = L$), and their mutual inductance is $M = L/2$. The two-loop digital voltage controller proposed in Figure 3 consists of a MACC [33] inner current programmed controller and a discrete-time PI compensator at the outer voltage feedback loop. Note, in Figure 3 we also represent two measuring approaches that allows to obtain different quantization levels of the measured output voltage error.

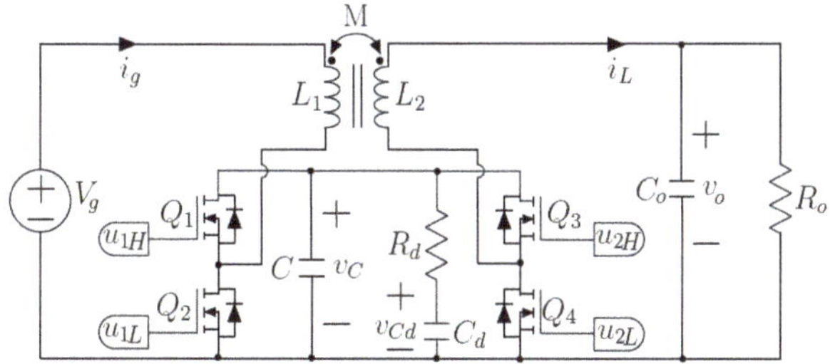

Figure 2. Power stage of a coupled-inductor buck-boost converter.

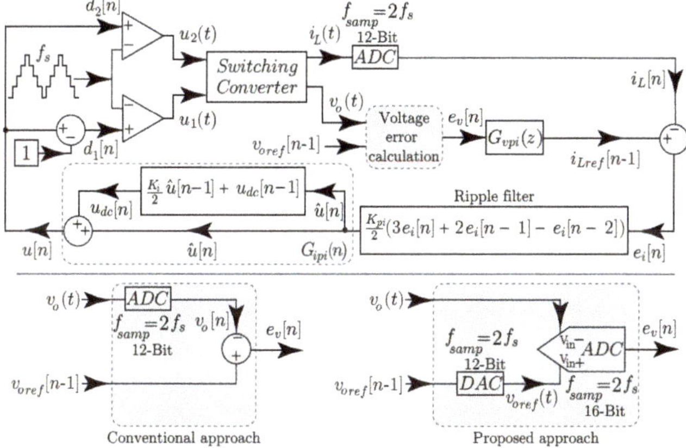

Figure 3. Block diagram of the digital controller for the voltage regulation of the buck-boost converter. Bottom Left: Conventional voltage error subcircuit. Bottom Right: Proposed improved approach subcircuit.

3.1. Multisampled Average Current Control (MACC)

The multisampled average current control for the bidirectional buck-boost converter was presented in [33]. The MACC stage generates the control variable (u) that is processed by a dual digital PWM to obtain the discrete control signals (u_1 and u_2) that activate the converter half-bridges. The external loop regulates the output voltage by providing the MACC with the output current reference through a discrete proportional-integral control transfer function $G_{vpi}(z)$, as it is seen in Figure 3. An important element of the MACC loop is the ripple filter processing the error between output current $i_L[n]$ and its desired reference $i_{Lref}[n-1]$. The ripple filter averages two consecutive samples per switching period ($f_{samp} = 2f_s$) of the output current error. This strategy eliminates the switching ripple in the current loop without significant phase loss [34].

The discrete-time ripple filter transfer function can be expressed as

$$\hat{u}[n] = \frac{K_{pi}}{2}\left(3e_i[n] + 2e_i[n-1] - e_i[n-2]\right). \tag{15}$$

The proportional gain can be written in terms of the output current waveform slopes as

$$K_{pi} = \frac{K_n}{(m_1 + m_2)T} \tag{16}$$

where the output current has a periodic triangular waveform with rising and falling current slopes m_1 and $-m_2$, respectively. The expression $m_1 + m_2$ is obtained for each converter operation mode, yielding

$$m_1 + m_2 = \begin{cases} \dfrac{Mv_o[n]}{L^2 - M^2} & \text{for boost mode} \\ \dfrac{LV_g}{L^2 - M^2} & \text{for buck mode.} \end{cases} \tag{17}$$

Parameter K_n has been adjusted to 0.35 to obtain a crossover frequency (CF) of approximately 11 kHz and a phase margin (PM) of 58° as in [33].

The digital PI compensator in the z-domain added to the current control loop has been implemented using forward-Euler method as follows,

$$G_{ipi}(z) = 1 + \frac{K_i}{2}\frac{1}{z-1} \tag{18}$$

Figure 3 shows the implementation of the discrete-time PI compensator, whose integral gain can be chosen as in [33].

3.2. Digital Proportional-Integral Voltage Control

A slower outer voltage loop providing current reference i_{Lref} is added to the inner current loop. The PI voltage controller is designed taken into account the value of the output filter capacitor (C_o) and the desired loop-gain crossover frequency (f_c). The transfer function of the PI voltage controller can be expressed in the z domain using the forward Euler method as

$$G_{vpi}(z) = K_{pv} + \frac{K_{iv}T_{samp}}{z-1}z^{-1} \tag{19}$$

where $K_{pv} = C_o 2\pi f_c$, $K_{iv} = K_{pv}/T_i$, and T_{samp} is the sample period ($1/f_{samp}$). Therefore, the bandwidth of the voltage loop depends on the proportional coefficient (K_{pv}), while the phase margin (PM) is adjusted to be greater than 50° adjusting K_{pv} after setting $T_i = 10/(2\pi f_c)$ for the integral coefficient (K_{iv}). The forward-Euler method is used to find the recurrence equations for the discrete-time PI controller as

$$\begin{aligned} i_{Lp}[n] &= K_{pv}e_v[n] \\ i_{Li}[n] &= K_{iv}T_{samp}e_v[n] + i_{Li}[n-1] \\ i_{Lref}[n] &= i_{Lp}[n] + i_{Li}[n]. \end{aligned} \tag{20}$$

4. Simulation and Experimental Results

The set-up used to carried out the different experiments with the MACC-based two-loop digital control is shown in Figure 4. It is composed of a 400 V 1.6 kW buck-boost prototype converter with the parameters described in Table 1 and the TMS320F28377S DSC. The design of the buck-boost converter is presented in [32].

The tests were carried out changing quantization values and controller parameters, as described next. Test 1 has been done using the conventional voltage error approach shown in Figure 3 using a 12 bit ADC to take the samples of output current and voltage. The external loop compensator is designed to obtain a cross-over frequency of $f_c = 4$ kHz. Test 2 also corresponds to the conventional voltage error measurement approach used in Test 1 but tuning $f_c = 2$ kHz. In order to reduce the voltage error quantization value q_v,

Test 3 has been carried out with the proposed voltage error block shown in Figure 3, using a 16 bit ADC in differential mode and $f_c = 4$ kHz for the external closed loop. In Test 4, the ADC quantization level of the output current and voltage is increased, scaling ADC resolution to 8 and 11 bits respectively for the current and voltage sampled values [35], and using the conventional approach error voltage block in Figure 3. Loop gains of the external control loop for the last test are selected to obtain $f_c = 4$ kHz.

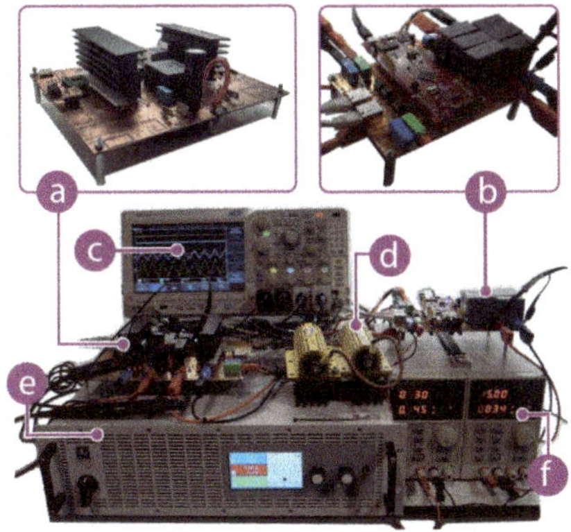

Figure 4. Experimental set-up of the buck-boost voltage regulator: (**a**) coupled-inductor buck-boost power stage, (**b**) digital signal controller with output capacitor $C_o = 28$ µF, (**c**) oscilloscope, (**d**) constant resistive load $R_o = 200$ Ω (**e**) input dc power supply, and (**f**) auxiliary power supply for DSC and MOSFET drivers.

Table 1. Parameters for the buck-boost setup.

Converter Parameters	Value
Input voltage V_g	200–400 V
Output voltage V_o	100–400 V
Rated power	1.6 kW
Switching frequency $f_s = 1/T$	100 kHz
Output capacitor C_o	28 µF
Intermediate capacitor C	1.32 µF
Mutual inductance $M = L_m$	135 µH
Self inductances $L_1 = L_2$	270 µH
Damping network $R_d C_d$	5 Ω, 20 µF
Load resistor R_o	200 Ω

Figure 5 shows simulated waveforms of output current reference i_{Lref}, variable control u, and voltage error e_v when the converter operates in steady-state with $V_g = 200$ V and $v_o = 300$ V.

A summary of the tests and evaluations of the fulfillment of the stability conditions for each loop, obtained by replacing the parameters of Tables 2 and 3 in the restrictions (11) and (14), are shown in Table 4. Figure 5a shows quantization-induced perturbations (QIP) in all signals for the Test 1, when neither $q_{DPWM}/q_i < K_{pi}$ for the restriction (14) of the inner current loop nor $K_{iv}T < q_i/q_v$ of the external loop are fulfilled. In Test 2, although condition (14) is not fulfilled, QIP are reduced for the output current reference as can be seen in Figure 5b. The condition for the external loop (11) is satisfied due to the reduction of the gains K_{iv} and K_{pv}, but, as the cross-over frequency depends on the proportional gain K_{pv}, the loop bandwidth is reduced. The ADC quantization level of the output voltage quantization q_v is reduced in

Test 3, where the condition for the external loop is fulfilled with a wide bandwidth, and the effects of QIP on the output current reference are significantly reduced.

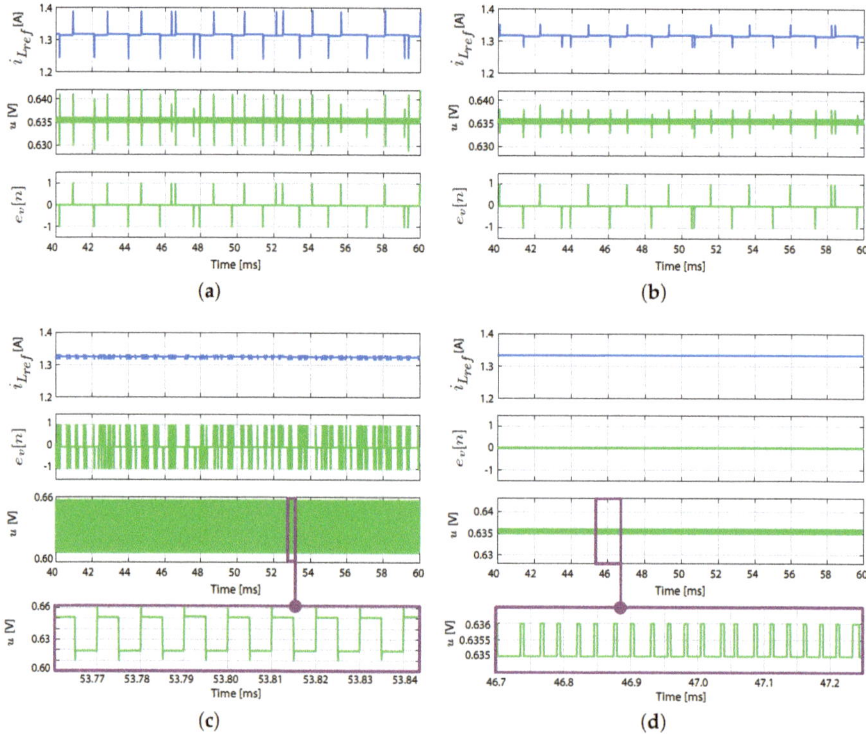

Figure 5. Simulation of output current reference i_{Lref}, signal control u, and voltage error e_v with the converter operating in steady-state: (**a**) Test 1; (**b**) Test 2; (**c**) Test 3; (**d**) Test 4.

Table 2. Analog parameters controller design.

External control parameters for $f_c = 4\,\text{kHz}$	Value
K_{pv}	$0.7\,\dfrac{A}{Vs}$
$K_{iv}T$	$0.07\,\dfrac{A}{V}$
External control parameters for $f_c = 2\,\text{kHz}$	**Value**
K_{pv}	$0.35\,\dfrac{A}{Vs}$
$K_{iv}T$	$0.035\,\dfrac{A}{V}$
Inner control parameters	**Value**
K_{pi}	$47\,\dfrac{1}{kAs}$
$K_{ii}T$	$4.7\,\dfrac{1}{kA}$

Table 3. ADC quantization parameters controller design.

Quantization Tests 1 and 2	Value
v_o ADC quantization level q_v	0.11 V
i_L ADC quantization level q_i	5.86 mA
DPWM quantization q_{DPWM}	0.002

Quantization Test 3	Value
v_o ADC quantization level q_v	0.013 V
i_L quantization level q_i	5.86 mA
DPWM quantization q_{DPWM}	0.002

Quantization Test 4	Value
quantization level q_v	0.22 V
quantization level q_i	93.75 mA
DPWM quantization q_{DPWM}	0.002

Table 4. Summary of the tests.

Test	$e_v[n]$ Calculation	f_c [kHz]	Condition (14)	Condition (11)
Test 1	Conventional	4	X	X
Test 2	Conventional	2	X	✓
Test 3	Proposed	4	X	✓
Test 4	Conventional	4	✓	✓

To compare the bandwidths and stability margins provided by each of the tests, the corresponding Bode plots of the voltage loop-gains for the converter operating in boost mode are provided in Figure 6, being the loop gain frequency response in a switched converter a powerful tool commonly used for the design of the controllers used in the control stage [36]. It is important to note that the results for the Test 1 have not been included in the frequency response analysis previously described. This test does not present a stable inner loop regulation which is evidenced by the presence of high current peak perturbations. This peak would be destructive for the converter if an experimental frequency response analysis is performed (please see the temporary experimental results presented below). Experimental plots in Figure 6b show that the Tests 3 and 4 with $K_{pv} = 0.7$ provide a CF of 4 kHz and a PM of 52°, while Test 2 with $K_{pv} = 0.35$ A/Vs, with smaller quantization perturbations (see Figure 5b), yields a CF = 2.16 kHz and PM = 59.58°.

Simulation tests were done using two voltage error measurement approaches shown in Figure 3. The conventional approach on the left side uses an ADC for sampling the voltage and then computes the voltage error. On the right side, the proposed method quantizes the voltage error by using an ADC in differential mode. Previously, it produces an analog voltage reference from the digital one. Through this last approach, it is possible to increase the error voltage resolution. The values of the analog control gains for the inner and for the external loop at different cross-over frequencies are shown in Table 2.

The proportional gain for the outer control is selected using the expression $K_{pv} = C_o 2\pi f_c$ for different crossover frequencies ($f_c = 2$ KHz and $f_c = 4$ kHz) and with $C_o = 28$ μF. Then, the proportional gain for the inner control is adjusted using the expression (16) with $K_n = 0.35$, employing the Equation (17) for boost mode with an output voltage of $V_o = 300$ V in Equation (16). The statement $K_i T = K_p/10$ ensures to obtain a PM greater than 50° (see Figure 6), therefore $K_{iv} T = K_{pv}/10$ for the outer loop and $K_{ii} T = K_{pi}/10$ for the inner loop. The parameters for the different tests are listed in Table 3.

Nonetheless, the condition for the inner loop is not fulfilled in this test, therefore the control variable u and voltage error e_v are not free of QIP effects as it is shown in Figure 5c. The simulated results of Test 4, in which both stability conditions are satisfied, are shown in Figure 5d, where the QIP perturbations have disappeared from all signals. It is also noticeable that, in comparison with previous tests, the control effort has been also reduced, which is indicated by the small amplitude of control variable u.

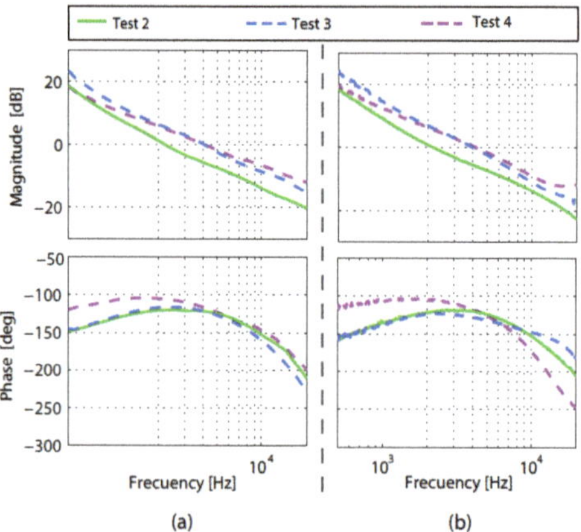

Figure 6. Voltage loop-gain Bode plots: (**a**) simulated, (**b**) experimental.

These values are in good agreement with the simulated results in Figure 6a. Despite the fact that Tests 2 and 3 do not satisfy all conditions, it is possible to operate the converter with these designs, obtaining a wider bandwidth using the proposed approach seen in Figure 3 for the Test 3.

Additional experiments and simulations have been performed to observe the current waveforms during start-up together with about 12 ms of steady-state regimes. Figure 7 depicts waveforms of input (i_g) and output (i_L) currents, as well as input (v_o) and output (V_g) voltages in the same cases previously shown in Figure 5.

In Figure 7, waveforms of the experimental results show higher QIP in relation to the simulation results due to noise in the experimental tests. In the same way, Figure 7a,b corresponds to Test 1 using the conventional approach, where the DPWM was configured for 8.96-bit. Output voltage and current ADC resolution are set to 12 bits, and the input ADC input voltage range goes from 0 V to 3 V. Figure 7a,b shows the simulated and experimental results when the proportional gain of the voltage loop is selected as $K_{pv} = 0.7$. In this case, the current waveforms present perturbations with high current overshoot and undershoot values. Results when the gain K_{pv} is reduced to the more conservative value of 0.35 in Test 2 are plotted in Figure 7c,d, showing that limiting the voltage loop bandwidth using the conventional error-calculation method reduces QIP in both currents improving the closed-loop stability. The current waveforms in Test 3 with the proposed improved error measurement approach in Figure 7e,f, show that there are no significant current perturbations when the proportional gain is again selected to $K_{pv} = 0.7$, so that a wide bandwidth voltage loop is obtained with a phase margin larger than 50°. In this case, the DPWM has been configured with 8.96-bit resolution, while the ADCs sampling the output current and the voltage error have been configured with resolutions of 12-bit and 16-bit differential mode, respectively. Figure 7g,h shows the simulated and experimental results for the Test 4, when the proportional gain is $K_{pv} = 0.7$ and both conditions (11) and (14) are fulfilled. Time domain current waveforms of the prototype for the different tests when it works in boost mode are shown in Figure 8. Figure 8a current waveforms for Test 1 present high current undershoot because conditions (11) and (14) are not fulfilled. The rest of test results show that the QIP is reduced, obtaining better results for the Test 4 with a $f_c = 4$ kHz (Figure 8d). It is important to remark that fulfilling condition (11) for the external loop is enough to reduce QIP and

LCO at the current waveforms in steady-state when the control of the converter is a two-loop with an integrator due to the external loop not cause induced perturbations in the internal loop.

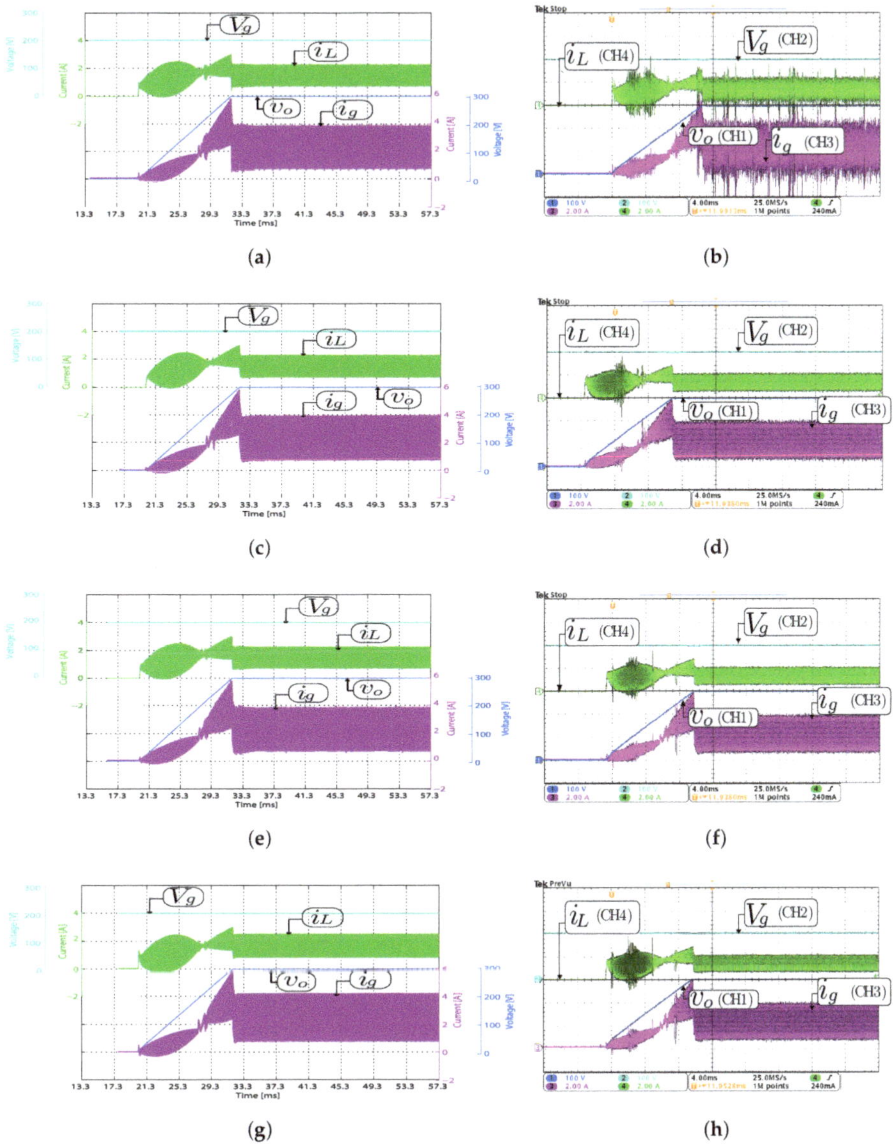

Figure 7. Simulated (**a,c,e,g**) and experimental (**b,d,f,h**) start-up waveforms: (**a,b**) Test 1, (**c,d**) Test 2, (**e,f**) Test 3, and (**g,h**) Test 4 ($V_g = 200$ V, $v_o = 300$ V, and $R_o = 200\ \Omega$). CH1: v_o (100 V/div). CH2: V_g (100 V/div) CH3: i_g (2 A/div), CH4: i_L (2 A/div).

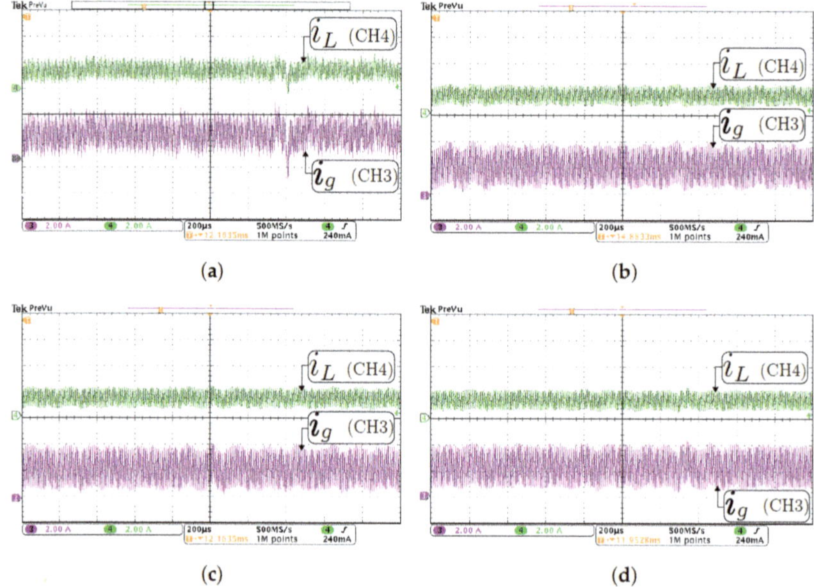

Figure 8. Time domain waveforms of i_g and i_L: (**a**) Test 1, (**b**) Test 2, (**c**) Test 3, (**d**) Test 4. CH3: i_g (2 A/div), CH4: i_L (2 A/div), and time base of 200 µs.

5. Conclusions

Limit cycle oscillations conditions due to quantization-induced perturbation in a digital two-loop current controlled converter are presented and analyzed in this paper. LCO conditions includes both loops ADCs quantization, DPWM quantization and gains of the to control laws to show the undesired quantization effects in a two-loop digital voltage regulator of a dc-dc converter with an integrator at its output. Simulation and experimental results, obtained after developing different tests on a 400 V 1.6 kW coupled-inductor buck-boost purpose-built prototype, validate that the current waveforms present perturbations when these conditions are not fulfilled. These tests also demonstrate that fulfilling the condition for the external loop is enough to reduce the quantization induced perturbations. Nevertheless, the comparison of test results suggests that fulfilling conditions for both loops is the best option to avoid LCO and QIP in the system variables. The work presents a useful guidance for the design of the PI digital controllers in DC-DC converters, in order to improve significantly the dynamic responses, increasing the voltage loop bandwidth without ADC quantization effects.

Author Contributions: Conceptualization, J.C. and E.V.-I.; methodology, C.G.-C.; software, C.R.; validation, C.G.-C., C.R., and R.G.; formal analysis, J.C.; investigation, C.G.-C. resources, C.R.; data curation, C.R.; writing—original draft preparation, C.G.-C. and R.G.; writing—review and editing, C.R., R.G., and E.V.-I.; visualization, C.R. and C.G.-C.; supervision, J.C.; project administration, E.V.-I.; funding acquisition, E.V.-I. All authors have read and agreed to the published version of the manuscript.

Funding: This work was supported by the Spanish Agencia Estatal de Investigación (AEI) and the Fondo Europeo de Desarrollo Regional (FEDER) under research projects DPI2016-80491-R (AEI/FEDER, UE) and DPI2017-84572-C2-1-R (AEI/FEDER, UE). The work was also supported by the the Chilean Government under Project CONICYT/FONDECYT 1191680 and by SERC Chile (CONICYT/FONDAP/15110019).

Conflicts of Interest: The authors declare no conflicts of interest.

Abbreviations

The following abbreviations are used in this manuscript.

ADC	Analog to digital converter
QIP	Quantization-induced-perturbation
MACC	Multrisampled average current control
DPWM	Digitally controlled pulse width modulation
DSC	Digital signal controller
LCO	Limit-cycle oscillations
PM	Phase margin
PI	Proportional-Integral
CF	Crossover frequency

References

1. Hayes, B. Nonlinear Dynamics of DC-DC Converters. Ph.D. Thesis, Dublin City University, Dublin, Ireland, 2016.
2. Tsai, C.H.; Yang, C.H.; Wu, J.C. A digitally controlled switching regulator with reduced conductive EMI spectra. *IEEE Trans. Ind. Electron.* **2012**, *60*, 3938–3947. [CrossRef]
3. Roy, S.; Murphree, R.C.; Abbasi, A.; Rahman, A.; Ahmed, S.; Gattis, J.A.; Francis, A.M.; Holmes, J.; Mantooth, H.A.; Di, J. A SiC CMOS digitally controlled PWM generator for high-temperature applications. *IEEE Trans. Ind. Electron.* **2017**, *64*, 8364–8372. [CrossRef]
4. Corradini, L.; Mattavelli, P. Modeling of multisampled pulse width modulators for digitally controlled DC–DC converters. *IEEE Trans. Power Electron.* **2008**, *23*, 1839–1847. [CrossRef]
5. Yan, X.; Shu, Z.; Sharkh, S.M.; Wu, Z.G.; Chen, M.Z. Sampled-Data Control With Adjustable Switching Frequency for DC–DC Converters. *IEEE Trans. Ind. Electron.* **2018**, *66*, 8060–8071. [CrossRef]
6. Neacsu, D.O.; Sirbu, A. Design of a LQR-Based Boost Converter Controller for Energy Savings. *IEEE Trans. Ind. Electron.* **2019**, *67*, 5379–5388. [CrossRef]
7. Darcy Gnana Jegha, A.; Subathra, M.; Manoj Kumar, N.; Subramaniam, U.; Padmanaban, S. A High Gain DC-DC Converter with Grey Wolf Optimizer Based MPPT Algorithm for PV Fed BLDC Motor Drive. *Appl. Sci.* **2020**, *10*, 2797. [CrossRef]
8. Zhou, M.; Yang, M.; Wu, X.; Fu, J. Research on Composite Control Strategy of Quasi-Z-Source DC–DC Converter for Fuel Cell Vehicles. *Appl. Sci.* **2019**, *9*, 3309. [CrossRef]
9. Hoyos, F.E.; Candelo-Becerra, J.E.; Hoyos Velasco, C.I. Application of Zero Average Dynamics and Fixed Point Induction Control Techniques to Control the Speed of a DC Motor with a Buck Converter. *Appl. Sci.* **2020**, *10*, 1807. [CrossRef]
10. Peng, H.; Prodic, A.; Alarcón, E.; Maksimovic, D. Modeling of quantization effects in digitally controlled dc–dc converters. *IEEE Trans. Power Electron.* **2007**, *22*, 208–215. [CrossRef]
11. Peterchev, A.V.; Sanders, S.R. Quantization resolution and limit cycling in digitally controlled PWM converters. *IEEE Trans. Power Electron.* **2003**, *18*, 301–308. [CrossRef]
12. Zhao, Z.; Prodic, A. Non-zero error method for improving output voltage regulation of low-resolution digital controllers for SMPS. In Proceedings of the 2008 Twenty-Third Annual IEEE Applied Power Electronics Conference and Exposition, Austin, TX, USA, 24–28 February 2008; pp. 1106–1110.
13. Syed, A.E.; Patra, A. Saturation generated oscillations in voltage-mode digital control of DC–DC converters. *IEEE Trans. Power Electron.* **2016**, *31*, 4549–4564. [CrossRef]
14. Syed, A.E.; Patra, A. Dynamic ADC-quantization for oscillation-free performance of digitally controlled converters. In Proceedings of the 2017 IEEE International Symposium on Circuits and Systems (ISCAS), Baltimore, MD, USA, 28–31 May 2017; pp. 1–4.
15. Khodamoradi, A.; Liu, G.; Mattavelli, P.; Messo, T. Simultaneous Identification of Multiple Control Loops in DC Microgrid Power Converters. *IEEE Trans. Ind. Electron.* **2019**, *67*, 10641–10651. [CrossRef]

16. Khatun, K.; Ratnam, V.V.; Rathore, A.K.; Narasimharaju, B.L. Small-Signal Analysis and Control of Soft-Switching Naturally Clamped Snubberless Current-Fed Half-Bridge DC/DC Converter. *Appl. Sci.* **2020**, *10*, 6130. [CrossRef]
17. Honkanen, J.; Hannonen, J.; Korhonen, J.; Nevaranta, N.; Silventoinen, P. Nonlinear PI-Control Approach for Improving the DC-Link Voltage Control Performance of a Power-Factor-Corrected System. *IEEE Trans. Ind. Electron.* **2018**, *66*, 5456–5464. [CrossRef]
18. Min, R.; Tong, Q.; Zhang, Q.; Zou, X.; Yu, K.; Liu, Z. Digital sensorless current mode control based on charge balance principle and dual current error compensation for DC–DC converters in DCM. *IEEE Trans. Ind. Electron.* **2015**, *63*, 155–166. [CrossRef]
19. Veerachary, M. Two-loop controlled buck–SEPIC converter for input source power management. *IEEE Trans. Ind. Electron.* **2011**, *59*, 4075–4087. [CrossRef]
20. Vidal-Idiarte, E.; Carrejo, C.E.; Calvente, J.; Martínez-Salamero, L. Two-loop digital sliding mode control of DC–DC power converters based on predictive interpolation. *IEEE Trans. Ind. Electron.* **2010**, *58*, 2491–2501. [CrossRef]
21. Dutta, S.; Hazra, S.; Bhattacharya, S. A digital predictive current-mode controller for a single-phase high-frequency transformer-isolated dual-active bridge DC-to-DC converter. *IEEE Trans. Ind. Electron.* **2016**, *63*, 5943–5952. [CrossRef]
22. Stefanutti, W.; Monica, E.D.; Tedeschi, E.; Mattavelli, P.; Saggini, S. Reduction of quantization effects in digitally controlled dc-dc converters using inductor current estimation. In Proceedings of the 2006 37th IEEE Power Electronics Specialists Conference, Jeju, Korea, 18–22 June 2006; pp. 1–7.
23. Chattopadhyay, S. Analysis of limit cycle oscillations in digital current-mode control. In Proceedings of the Twenty-First Annual IEEE Applied Power Electronics Conference and Exposition (APEC'06), Dallas, TX, USA, 19–23 March 2006; p. 7.
24. Kapat, S. Selectively sampled subharmonic-free digital current mode control using direct duty control. *IEEE Trans. Circuits Syst. II Express Briefs* **2015**, *62*, 311–315. [CrossRef]
25. Sajitha, G.; Kurian, T. Reduction in limit cycle oscillation and conducted electromagnetic emissions by switching frequency adjustment in digitally controlled DC-DC converters. *EPE J.* **2016**, *26*, 96–103.
26. Hu, H.; Yousefzadeh, V.; Maksimovic, D. Nonuniform A/D Quantization for ImprovedDynamic Responses of Digitally ControlledDC–DC Converters. *IEEE Trans. Power Electron.* **2008**, *23*, 1998–2005.
27. Sanchez, A.; de Castro, A.; López, V.M.; Azcondo, F.J.; Garrido, J. Single ADC digital PFC controller using precalculated duty cycles. *IEEE Trans. Power Electron.* **2013**, *29*, 996–1005. [CrossRef]
28. Corradini, L.; Maksimovic, D.; Mattavelli, P.; Zane, R. *Digital Control of High-Frequency Switched-Mode Power Converters*; John Wiley & Sons: Hoboken, NJ, USA, 2015; Volume 48.
29. Peng, H.; Maksimovic, D. Digital current-mode controller for DC-DC converters. In Proceedings of the Twentieth Annual IEEE Applied Power Electronics Conference and Exposition, 2005, APEC 2005, Austin, TX, USA, 6–10 March 2005; Volume 2, pp. 899–905.
30. Restrepo, C.; Calvente, J.; Cid-Pastor, A.; El Aroudi, A.; Giral, R. A noninverting buck–boost DC–DC switching converter with high efficiency and wide bandwidth. *IEEE Trans. Power Electron.* **2011**, *26*, 2490–2503. [CrossRef]
31. González-Castaño, C.; Vidal-Idiarte, E.; Calvente, J. Design of a bidirectional DC/DC converter with coupled inductor for an electric vehicle application. In Proceedings of the 2017 IEEE 26th International Symposium on Industrial Electronics (ISIE), Edinburgh, UK, 19–21 June 2017; pp. 688–693.
32. González-Castaño, C.; Restrepo, C.; Giral, R.; García-Amoros, J.; Vidal-Idiarte, E.; Calvente, J. Coupled inductors design of the bidirectional non-inverting buck–boost converter for high-voltage applications. *IET Power Electron.* **2020**. [CrossRef]
33. Restrepo, C.; Konjedic, T.; Flores-Bahamonde, F.; Vidal-Idiarte, E.; Calvente, J.; Giral, R. Multisampled Digital Average Current Controls of the Versatile Buck-Boost Converter. *IEEE J. Emerg. Sel. Top. Power Electron.* **2018**, *7*, 879–890. [CrossRef]
34. Corradini, L.; Stefanutti, W.; Mattavelli, P. Analysis of multisampled current control for active filters. *IEEE Trans. Ind. Appl.* **2008**, *44*, 1785–1794. [CrossRef]

35. Lin, Y.C.; Chen, D.; Wang, Y.T.; Chang, W.H. A novel loop gain-adjusting application using LSB tuning for digitally controlled DC–DC power converters. *IEEE Trans. Ind. Electron.* **2011**, *59*, 904–911. [CrossRef]
36. Gonzalez-Espin, F.; Figueres, E.; Garcerá, G.; González-Medina, R.; Pascual, M. Measurement of the loop gain frequency response of digitally controlled power converters. *IEEE Trans. Ind. Electron.* **2009**, *57*, 2785–2796. [CrossRef]

Publisher's Note: MDPI stays neutral with regard to jurisdictional claims in published maps and institutional affiliations.

 *applied sciences*

Article

Analysis of Subharmonic Oscillation and Slope Compensation for a Differential Boost Inverter

Abdelali El Aroudi [1,*] , Mohamed Al-Numay [2] , Reham Haroun [1] and Meng Huang [3]

[1] Department of Electronics, Electrical Engineering and Automatic Control, Universitat Rovira i Virgili, 43002 Tarragona, Spain; reham.haroun@urv.cat
[2] Electrical Engineering Department, King Saud University, 11451 Riyadh, Saudi Arabia; alnumay@ksu.edu.sa
[3] School of Electrical Engineering, Wuhan University, Wuhan 430072, China; meng.huang@whu.edu.cn
* Correspondence: abdelali.elaroudi@urv.cat; Tel.: +34-977-558-522

Received: 10 July 2020; Accepted: 11 August 2020; Published: 13 August 2020

Abstract: This paper focuses on the steady-behavior of a differential boost inverter used for generating a sinewave AC voltage in rural areas. The analysis of its dynamics will be performed using an accurate approach based on discrete time models and Floquet theory and adopting a quasi-static approximation. In particular, the undesired subharmonic oscillation exhibited by the inverter will be analyzed and its boundary in the parameter space will be predicted and delimited. Combining analytical expressions and computational procedures to determine the quasi-static duty cycle, subharmonic oscillation is accurately predicted. It is found that subharmonic oscillation takes place at critical values of the sinewave voltage reference cycle, which can cause distortion to the input current and degrade the harmonic content of the output voltage. The results provide useful information for the design of the boost inverter to avoid distortion caused by subharmonic oscillation. Namely, the minimum value of the compensation slope and the maximum proportional gain of the AC output voltage controller guaranteeing a pure sinewave voltage and clean inductor current during the entire AC cycle will be determined. Numerical simulations performed on the switched model implemented using PSIM© software confirm the theoretical predictions.

Keywords: differential boost inverter; current mode control; nonlinear behavior; subharmonic oscillation; slope compensation

---

1. Introduction

DC-AC inverters find widespread usage in many residential, industrial and military applications. With the ever-increasing development of the renewable energy technology, DC-AC inverters have become one of the most attractive and viable solutions to the power conversion problem. They are extensively used and play key roles in various actual applications of power electronics technologies for renewable energy sources [1–3]. They are also used in motor drive [4,5] and DSTATCOM applications [6] as well as in many uninterruptible power supply system applications such as plant facilities and factories, medical equipments and centers in hospitals, airline computer and communication systems in server farms and web hosting sites [7]. One of the important tasks in the design of DC-AC inverters is the control loop implementation which must ensure a system free from any kind of instabilities. However, it is well known that this aim is difficult to be achieved for all values of system parameters and that many undesired nonlinear phenomena can arise in these kinds of indispensable parts of modern and emerging energy systems. These phenomena can significantly jeopardize the system performance and can cause serious consequences on its reliability.

Therefore, understanding these nonlinear phenomena, their analysis, prediction and control have increasingly become of great concern of many researchers all over the world [8–22]. The major part of the analytical results on subharmonic oscillation in power electronics converters has been achieved for DC-DC converters [23–40]. DC-AC inverters are more difficult to deal with, since their dynamics is governed by two vastly different frequencies, namely the high switching frequency and the low frequency of the output voltage reference sinewave.

For reliable and desirable operation, the stability of the system must be guaranteed for the whole range of its parameters. In [13], the dynamics behavior of an H-bridge under a digital Current Mode Control (CMC) was investigated by using a one dimensional discrete time model. Different dynamical behaviors for the system were revealed by varying the proportional gain of the current controller. In [14] a similar approach was applied and it was demonstrated that different types of bifurcations (instabilities) can take place such as period doubling leading to Subharmonic Oscillation (SO) and border collision bifurcations leading directly to chaotic behavior.

Using the quasi-static approximation, in [15] the slow-scale and fast-scale instabilities in a voltage-mode controlled H-bridge inverter are reported and analyzed using an averaged model and a discrete-time model respectively. It is well known that conventional averaged model cannot predict the fast-scale instability and for that the discrete-time model must be used. A closed form discrete time model was used in [16] to predict both the slow-scale and the fast-scale instabilities in an H-bridge inverter demonstrating that the system may undergo instability phenomenon when the proportional gain of the voltage controller is increased. In an H-bridge digital-controlled grid-connected inverter system, bifurcation behavior was investigated and loss of system stability was shown by increasing the current controller gain [19] and it was shown that in this system only slow scale instability may take place leading to low-frequency oscillation. The same system, but with double edge modulation, has been studied in [9] using an analytical closed-form expression for predicting a period doubling phenomenon.

Single-stage grid-connected DC-AC conversion systems with boosting voltage capability have recently attracted the attention of many researchers. Single-stage structures of inverters not only perform DC-AC conversion but also perform voltage boosting. Moreover, differential inverter topologies seem to prevail in price and size due to the utilization of small passive elements of DC-DC converters hence improving the efficiency. In contrast to the conventional H-bridge inverter, the differential boost inverter is a flexible DC-AC inverter topology providing voltage step-up capability and could be a potential candidate for many DC-AC electrical energy conversion applications such as for power processing stage fuel-cell energy system [41,42], for high quality sine wave generation with a high oscillation frequency [43], for AC-module microinverters in PV systems such as in [44–46] among others.

In stand-alone operation mode, the load is directly supplied by the inverter. Single-phase H-bridge inverters are simple bidirectional converter topologies capable of handling both real and reactive power having their performance evaluated in terms of power quality and stability. Therefore, generating a high quality output voltage with low distortion and good voltage regulation is the main target. Other relevant performance metrics include disturbance rejection, transient response, and insensitivity to load and system parameter variations. These metrics can only be achieved with a design free from any kind of instability.

Since its introduction in [47], many studies have dealt with the control design of the differential boost inverter using different approaches and strategies [44,45,47–49]. The focus in most of the works published about this inverter is on the control design. However, the analysis of its nonlinear behavior has not been addressed in the past. Namely, SO has not been studied in this kind of inverter. Therefore the aim of this paper is to apply the Floquet theory for accurately predicting the onset of SO in a differential boost inverter. In contrast to existing works on predicting such a complex behavior in DC-AC inverters based mainly on numerical procedures, here both numerical and analytical approaches are combined to provide a comprehensive study of the systems dynamical behavior.

The prediction of this phenomenon is of high importance from both theoretical and practical points of view because it leads to an increase in the ripple of the currents and voltages and this has a harmful effect on the system performances since the overall losses become more significant. The power quality can also be jeopardized if SO is more pronounced since it can increase the THD and the current stress on the switches. Therefore, accurate modeling and stability analysis are necessary for exploring the dynamic behavior and predicting the stability boundaries of DC-AC inverters.

The remaining of this paper is organized as follows. In Section 2, the system dealt with in this study is described. In Section 3, the dynamic behavior of the system is explored revealing that the behavior of the system waveforms is phase-dependent. The system is shown to exhibit local instability phenomenon over a specific interval within the main sinusoidal cycle. The onset of the observed bubbling is associated to a SO phenomenon taking place at the fast switching scale. The mathematical modeling is addressed in Section 4 in the continuous-time domain. In order to analyze the observed phenomena in Section 3, Floquet theory is applied to the derived model in Section 5. Thereafter, in Section 6, the stability boundaries in terms of suitable parameters is reported. Finally, in Section 7 the results of the study are summarized.

2. Differential Boost Inverter under Two-Loop Control

The system under study in this paper consists of a differential boost inverter which is obtained by connecting two identical DC-DC boost converters in parallel supplied from a common electrical energy source and feeding a floating voltage load connected between the outputs of the two converters [47,50]. Its schematic diagram is shown in Figure 1. The current drawn by the input is shared properly between the two boost converters by the action of a CMC scheme using the difference between the two inductor currents, as will be detailed later. For that, two complementary control signals are considered to control the switches of the differential inverter.

Let us denote the two connected converters as Converter 1 with inductor L_1 and inductor current i_1 and Converter 2 with inductor L_2 and inductor current i_2. Both converters are controlled in a complementary way using CMC via single Pulse-Width Modulation (PWM) scheme so that Converter 2 is phase shifted $2\pi D$ with respect to Converter 1 at the switching time scale, D being the operating duty cycle. Namely, the difference between i_1 and i_2 (scaled by a sensing resistance r_s) is controlled using a conventional peak CMC by comparing the signal $r_s(i_1 - i_2)$ to the signal $r_s i_{ref}$. A periodic ramp signal v_{ramp} with amplitude V_M and period T is subtracted from $r_s i_{ref}$ for slope compensation. The comparison of the signal $r_s(i_1 - i_2)$ with the signal $r_s i_{ref} - v_{ramp}$ by using a comparator and a set-reset flip-flop generate the high and low values of the pulses driving the switches as shown in Figure 1 where the block diagram of the inner current control together with the outer voltage control are depicted.

The reference current for the difference between the two inductor currents is provided by an external voltage loop. The activation of the switches Q_1, Q_2, Q_3 and Q_4 is carried out as follows: the signal $r_s(i_1 - i_2)$ is connected to the non inverting pin of the comparator whereas the signal $r_s i_{ref} - v_{ramp}$ is applied to the inverting pin. The output of the comparator is applied to the reset input of a set-reset flip-flop and a periodic clock signal is connected to its set input, as shown in Figure 1, in such a way that the switch Q_2 and Q_4 are ON at the beginning of each switching cycle and are turned OFF whenever $r_s(i_1 - i_2) = r_s i_{ref} - v_{ramp}$. The state of the switches Q_1 and Q_3 are complementary to the switches Q_2 and Q_4 respectively.

To fulfill the requirements of the underlying electronic application, a DC-AC inverter has to produce a periodic sinewave-shaped output voltage under normal operational conditions. Let $v_{ref}(t)$ be the voltage reference that can be expressed as $v_{ref}(t) = V_{ref}\sin(2\pi f_g t) = V_{ref}\sin(\varphi)$, where $\varphi = 2\pi f_g t \in (0, 2\pi)$, V_{ref} is the peak value of the output voltage reference, ω_0 its angular frequency and φ its phase angle. In practical applications, the switching frequency is much higher than the AC output voltage frequency.

This condition is met in this paper and it allows the use of quasi-static approximation. The error voltage $v_{\mathrm{ref}} - v_o$ is the input signal to the voltage controller of which the task is to make the output voltage of the inverter an AC sinusoidal signal with zero DC component. Therefore, the load connected between the converters outputs will be subjected to an AC sinusoidal voltage with a zero DC component. This control strategy is different from the one used in most of the published works about this inverter topology such as [44,45,47] where the control is performed such that each boost converter generates a DC bias and an AC component. In the low frequency averaged sense, the AC component of each converter is out of phase regarding the other converter. The DC component is the same for both converters.

The voltage controller is conventionally a PI regulator aiming to make the load voltage v_o to accurately track the sinewave voltage reference v_{ref}. Its transfer function can be expressed as $H_{\mathrm{pi}}(s) = k_p(\tau s + 1)/(s\tau)$, where k_p is its proportional gain and τ is its time constant.

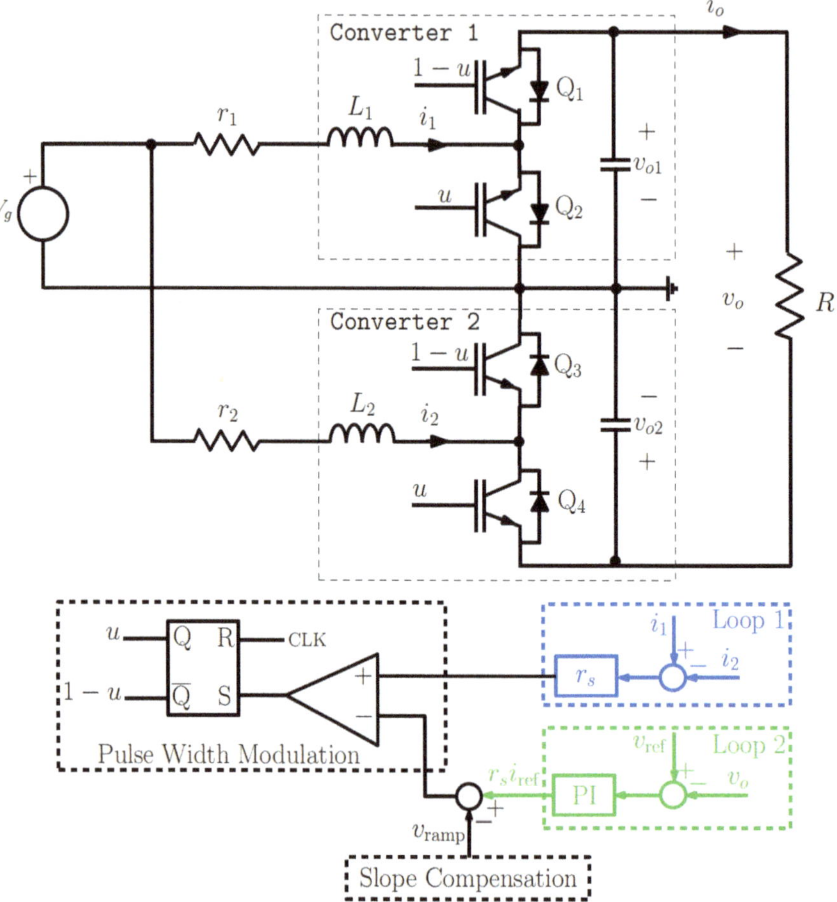

Figure 1. The differential boost inverter under two-loop control.

3. Behavior of the Differential Boost Inverter

The dynamical behavior of the boost inverter is explored in this section with the aim to gain insight on suitable ways of obtaining an appropriate model that can be used for its accurate stability analysis. The system is first studied through simulations using the full-order switched model of the inverter implemented using PSIM© software by varying suitable system parameters. The focus is first on system stability in terms of the time varying voltage reference. The fixed parameter values used for the rest of the study are reported in Table 1. Many time-domain waveforms have been computed to get a clear view of the system behavior and only representative results are shown below. The simulation is run for sufficiently long time to allow the system to reach its steady-state. The data obtained during time transient within the startup phase and during the transient regime of the regulation phase are fully eliminated. Only the last two cycles of the output voltage reference are plotted.

Table 1. The used parameters for the DC-AC differential boost inverter.

Parameter	Value
Inductance $L_1 = L_2$	100 µH
Resistance $r_1 = r_2$	0.1 Ω
Capacitance $C_1 = C_2$	22 µF
Input voltage v_g	200 V
Load resistance R	100 Ω
Time constant of the voltage controller τ	1 ms
PWM switching frequency f_s	100 kHz
RMS value of the reference voltage v_{ref}	230 V
Frequency of the reference voltage v_{ref}	50 Hz
Current sensor gain r_s	0.1 Ω

Figure 2 shows the system waveforms when the system is stable. The figure shows the time-domain waveforms of the reference voltage v_{ref} and the output voltage v_o, the capacitor voltages v_{o1} and v_{o2}, the inductor currents i_1 and i_2 and the control signal $r_s(i_1 - i_2)$ and the signal $r_s i_{ref} - v_{ramp}$. It is worth noting that the output voltage cannot be distinguished from its reference signal v_{ref} due to the practically zero amplitude and phase errors. Note also that the state variables and the control signal oscillate at two main frequencies, the switching frequency (100 kHz) and the reference voltage frequency (50 Hz). From a practical point of view, the output voltage is characterized by a low value of THD as required in any application.

As parameters are varied, the state variables undergo a sudden distortion by exhibiting SO at the fast switching scale as shown in Figure 3 for $k_p = 0.4$. This phenomenon takes place when the proportional gain k_p gradually increases and reaches a critical value close to 0.22. As shown in Figure 3, it can be observed that the inductor currents i_1 and i_2 exhibit SO leading to disrupting bubbling phenomenon of the waveforms. In particular, when $k_p \approx 0.22$, the fast-scale instability develops in all the state variables but it is more visible and pronounced in the inductor current waveforms i_1 and i_2 and their combination $r_s(i_1 - i_2)$. As stated before, such behavior manifests itself as a period-doubling phenomenon at the fast switching scale [9,51]. It can also be noticed in Figure 4 that the phenomenon already becomes visible in the capacitor voltages and the output voltage hence it can deteriorate the performance of the inverter and therefore its prediction is an important task from a practical point of view.

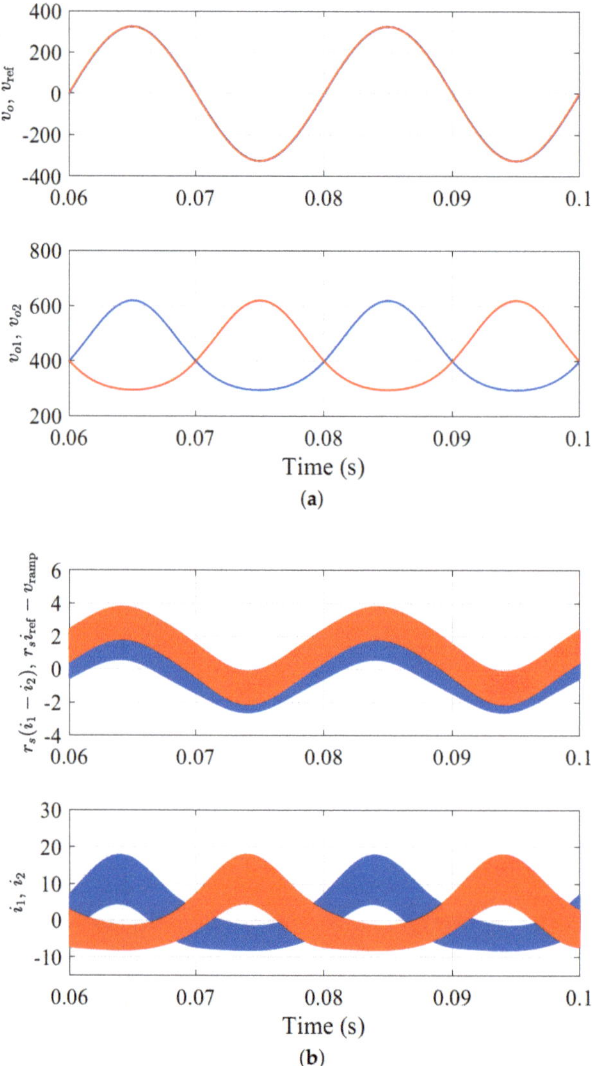

Figure 2. Steady-state response of boost inverter with $k_p = 0.2$ and $V_M = 2$ V. (**a**) Capacitor voltages v_{o1} and v_{o2}, output and reference voltages v_o and v_{ref}. (**b**) Inductor currents i_1 and i_2 and control signals $r_s(i_1 - i_2)$ and $r_s i_{ref} - v_{ramp}$. For each subplot, traces correspond to the shown voltages in [V] and currents in [A].

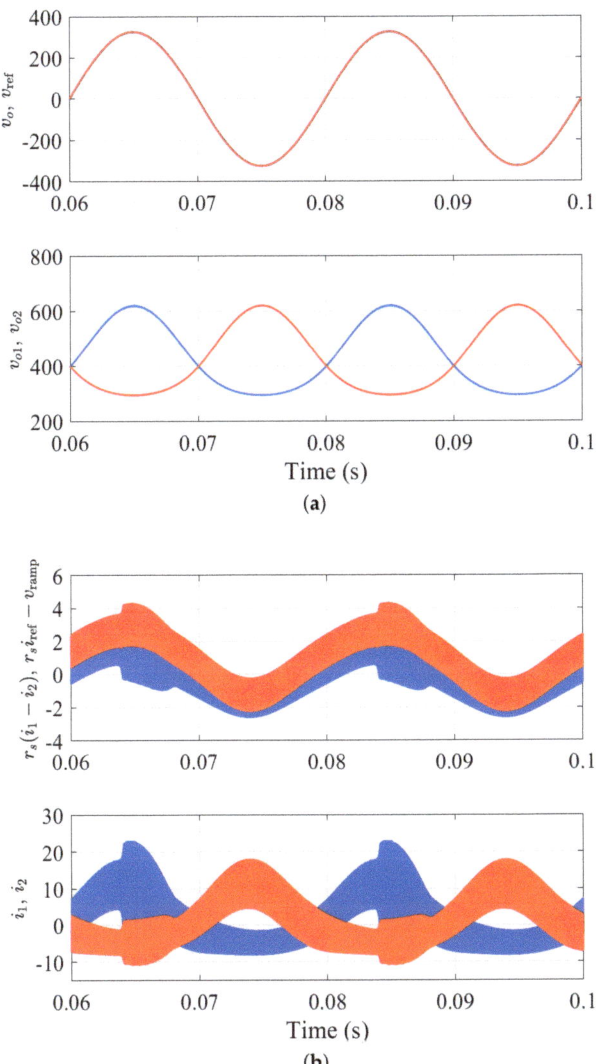

Figure 3. Steady-state response of boost inverter with $k_p = 0.4$ and $V_M = 2$ V. (**a**) Capacitor voltages v_{o1} and v_{o2}, output and reference voltages v_o and v_{ref}. (**b**) Inductor currents i_1 and i_2 and control signals $r_s(i_1 - i_2)$ and $r_s i_{\mathrm{ref}} - v_{\mathrm{ramp}}$. For each subplot, traces correspond to the shown voltages in [V] and currents in [A].

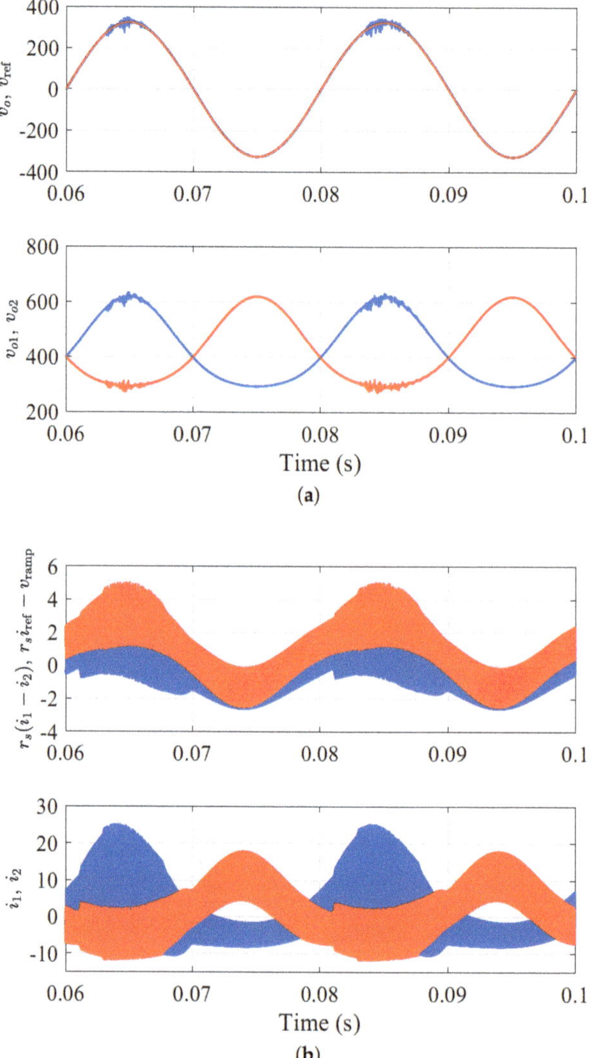

Figure 4. Steady-state response of boost inverter with $k_p = 0.8$ and $V_M = 2$ V. (**a**) Capacitor voltages v_{o1} and v_{o2}, output and reference voltages v_o and v_{ref}. (**b**) Inductor currents i_1 and i_2 and control signals $r_s(i_1 - i_2)$ and $r_s i_{\text{ref}} - v_{\text{ramp}}$. For each subplot, traces correspond to the shown voltages in [V] and currents in [A].

By carefully examining the waveforms, the following statements can be made:

- By increasing progressively the proportional gain and when this parameter reaches the critical value, SO oscillation starts first occurring in a very limited number of switching cycles during the first half cycle of the sinewave signal eventually in the neighborhood to the quarter of the cycle where the sinewave signal is maximum.
- The number of the switching cycles, during which SO is exhibited, gets larger and the fast-scale SO is more pronounced when the proportional gain k_p is increased.
- At the left and at the right of the maximum values during the same half cycle, one has the same values of quasi-steady-state duty cycles and therefore, theoretically, a perfect symmetry is expected in the critical phase angles at which SO takes place. However, an asymmetry can take place because the slope of the reference sinewave signal at the left of the peak point is positive while it is negative at the right side.
- The SO interval is repetitive from a sinewave cycle to the next one and the study of SO phenomenon can be restricted to one sinewave cycle in terms of the phase angle φ as a slowly varying parameter in the range $\varphi \in (0, 2\pi)$.
- Apparently, if SO is avoided for the first half cycle of the sinewave signal, it will also be avoided for the second half cycle. Therefore, the numerical and the analytical studies to be presented later will be restricted to the first half cycle of the sinewave signal for $\varphi \in (0, \pi)$, i.e., only within the duty cycle range $D \in (0.5, 1)$.

A powerful tool for clearly illustrating the SO phenomenon is by using the sampled waveforms. In order to clearly appreciate the change in the behavior of the system, sampled steady-state values of the state variables at time instants $t = nT$ ($n \in \mathbb{N}$) are obtained. Therefore, the state variables are sampled at every clock instant and then plotted in the time domain. A priori, any one of the state variables can be used for illustrating the behavior of the system. However, as observed in the previous time domain numerical simulations, SO is more pronounced in some state variables than others. An interesting and naturally sampled variable for which SO is well noticed is the duty cycle of the binary signal u.

Figure 5 shows the waveforms of the duty cycle $d(nT)$ ($n \in \mathbb{N}$) during one complete sinewave cycle for four different values of the proportional gain k_p. The duty cycle waveforms are plotted in terms of the phase angle within the interval $(0, 2\pi)$. For $k_p = 0.2$, the system exhibits a stable periodic regime in steady-state, the duty cycle does not present any disruption and its samples represent a clean and smooth waveform. When the SO regime starts taking place, one gets a different picture. For instance, for $k_p = 0.4$, it can be clearly seen that there is a certain phase interval within the first half of the sinewave cycle during which the duty cycle waveforms is disrupted. Namely, within the phase interval defined by two critical phase angles, two different branches of duty cycle values appear instead of one a kind of bubble emerges [18]. It can be observed that the onset of bubbling phenomenon depicted in Figure 5 is gradual. First, for a relatively small value of the parameter k_p, the cycle is smooth, then, for increasing k_p, it becomes disrupted in a small phase interval. Thereafter, as k_p is further increased, the interval (φ_1, φ_2) of φ during which SO takes place grows up as can be seen in Figure 5. If the proportional gain is further increased, this interval gets wider and the phenomenon usually spreads through the whole line cycle. Figure 5 also shows that successive period doubling inside the SO interval may also take place in the first half cycle where the voltage reference is positive, i.e., when $D > 0.5$. When k_p becomes even larger, the bubbles start appearing even in the second half cycle of voltage reference where $D < 0.5$. Therefore, even for $D < 0.5$, the voltage loop may have a destabilizing effect since when the proportional gain k_p is increased beyond a critical value $k_p \approx 0.8$, SO and the associated bubbling starts appearing for $D < 0.5$ and even in the presence of slope compensation. Therefore, the ramp slope needed for eliminating SO is larger than the one obtained when ignoring the effect of the voltage loop. This destabilizing effect of the voltage loop is similar to the one reported in [27] for the buck converter and in [24] for the boost converter. Similar behaviors have been obtained when other parameters such as the input voltage V_g or the inductance L are varied.

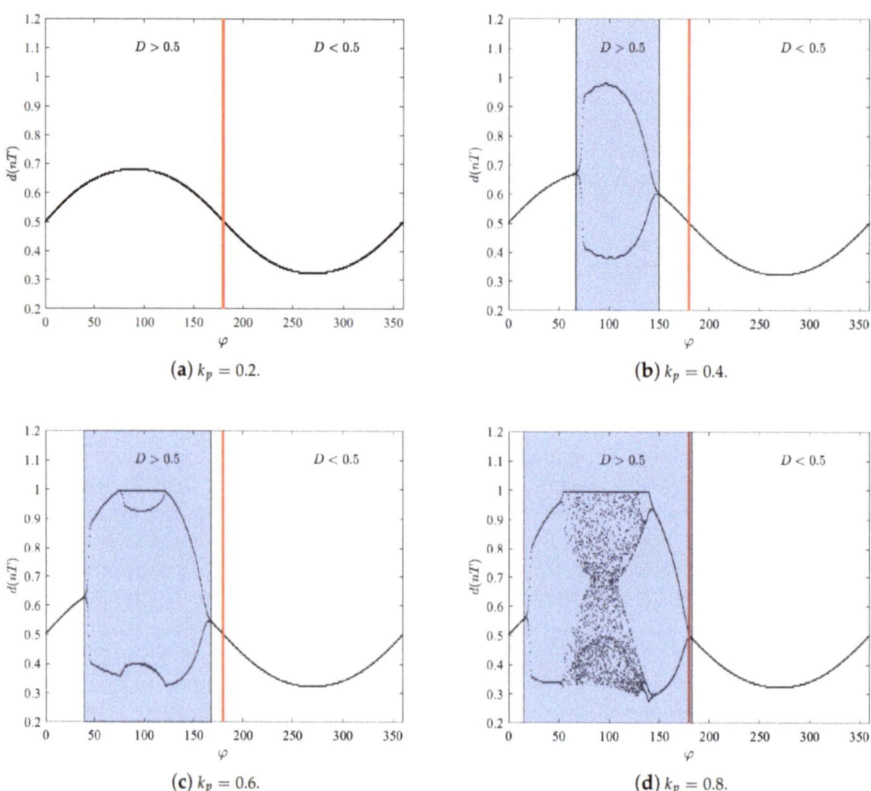

Figure 5. Waveforms of the duty cycle $d(nT)$ at steady-state operation in terms of the phase angle in [°] for different values of k_p and for $V_M = 2$ V.

4. Continuous-Time Modeling of the Differential Boost Inverter

4.1. Quasi-Steady-State Analysis

From the simulation results presented in the previous section, it has been observed that SO takes place when suitable parameters are varied. One of the widespread tools to analyze and to investigate this kind of nonlinear behavior is Floquet theory [52,53]. Considering the switched model of the system, one can identify possible periodic orbits, their stability as well as several other important aspects of the dynamical behavior. To apply this theory, the mathematical model is first derived. By applying KVL and KCL, the switched model of the differential boost inverter can be expressed as follows

$$\frac{di_1}{dt} = \frac{1}{L_1}(V_g - v_{o2}(1 - u)) - \frac{r_1}{L_1}i_1, \tag{1}$$

$$\frac{di_2}{dt} = \frac{1}{L_2}(V_g - uv_{o2}) - \frac{r_2}{L_2}i_2, \tag{2}$$

$$\frac{dv_{o1}}{dt} = \frac{1}{C_1}\left((1 - u)i_1 + \frac{v_{o1} - v_{o2}}{R}\right), \tag{3}$$

$$\frac{dv_{o2}}{dt} = \frac{1}{C_2}\left(ui_2 - \frac{v_{o1} - v_{o2}}{R}\right), \tag{4}$$

where L_1 and L_2 are the inductance of the inductors of the differential boost inverter with stray resistances r_1 and r_2 respectively, C_1 and C_2 are the capacitances of their capacitors. V_g is the DC input voltage and R is the AC load resistance. All other parameters appearing in (1)–(4) are shown in Figure 1. The quasi-steady-state average values of the state variables are related to the quasi-steady-state duty cycle D by the following expressions:

$$I_1 = \frac{V_g(2D - 1)}{RD(1 - D)^2}, \quad I_2 = -\frac{V_g(2D - 1)}{RD^2(1 - D)} \tag{5}$$

$$V_{o1} = \frac{V_g}{1 - D}, \quad V_{o2} = \frac{V_g}{D} \tag{6}$$

These expressions have been obtained by using the averaged model of the inverter within a switching period. Using (6) and the fact that $v_o = v_{o1} - v_{o2}$, the voltage gain of the differential boost inverter can be expressed as follows

$$M(D) := \frac{v_{ref}}{V_g} = \frac{2D - 1}{D(1 - D)} \tag{7}$$

The inverter gain $M(D)$ reaches its maximum value $M_{max} = V_{ref}/V_g$ when the voltage reference v_{ref} reaches its peak value V_{ref}. From the expression of $M(D)$, the steady-state value of the duty cycle can be derived and this can be expressed as follows

$$D(t) = \begin{cases} \dfrac{1}{2} - \dfrac{V_g}{v_{ref}} + \dfrac{\sqrt{4V_g^2 + v_{ref}^2}}{2v_{ref}} & \text{if } v_{ref}(t) > 0, \\[2ex] \dfrac{1}{2} - \dfrac{V_g}{v_{ref}} - \dfrac{\sqrt{4V_g^2 + v_{ref}^2}}{2v_{ref}} & \text{if } v_{ref}(t) < 0. \end{cases} \tag{8}$$

In terms of the phase angle φ, the quasi steady state duty cycle can be expressed as follows

$$D(\varphi) = \begin{cases} \dfrac{1}{2} - \dfrac{1}{M_{max}\sin(\varphi)} + \sqrt{M_{max} + \tfrac{1}{4}} & \text{if } \varphi \in (0, \pi), \\[2ex] \dfrac{1}{2} - \dfrac{1}{M_{max}\sin(\varphi)} - \sqrt{M_{max} + \tfrac{1}{4}} & \text{if } \varphi \in (\pi, 2\pi). \end{cases} \tag{9}$$

4.2. The State-Space Switched Model

Let $\mathbf{x} = (i_1, i_2, v_{o1}, v_{o2},)^\mathsf{T}$ be the vector of the state variables of the power stage of the inverter. The system can be described by a piecewise linear switched model as follows

$$\dot{x} = A_1 x + B_1 V_g, \quad \text{for} \quad u = 1, \tag{10}$$

$$\dot{x} = A_0 x + B_0 V_g, \quad \text{for} \quad u = 0, \tag{11}$$

$$\dot{v}_i = v_{\text{ref}} - (v_{o1} - v_{o2}) = v_{\text{ref}} - C^\mathsf{T} x \tag{12}$$

where $C^\mathsf{T} = (0 \quad 0 \quad 1 \quad -1)$ and $v_i := \int (v_{\text{ref}} - v_o)\,dt$ is the integral of the error signal $v_{\text{ref}} - v_o$. $A_0 \in \mathbb{R}^{4\times 4}$, $A_1 \in \mathbb{R}^{4\times 4}$, $B_0 \in \mathbb{R}^{4\times 1}$ and $B_1 \in \mathbb{R}^{4\times 1}$ are the system state matrices presented below. The variable v_i was deliberately separated from the rest of state variables to avoid matrix singularities appearing in the expressions of the system trajectories and their steady-state values at the switching time instants [25,26]. The matrices A_1, A_0, B_1 and B_0 are as follows:

$$A_1 = \begin{pmatrix} -\dfrac{r_1}{L_1} & 0 & 0 & 0 \\[6pt] 0 & -\dfrac{r_2}{L_2} & 0 & -\dfrac{1}{L_2} \\[6pt] 0 & 0 & -\dfrac{1}{RC_1} & \dfrac{1}{RC_1} \\[6pt] 0 & \dfrac{1}{C_2} & \dfrac{1}{RC_2} & -\dfrac{1}{RC_2} \end{pmatrix}, \quad B_1 = \begin{pmatrix} \dfrac{1}{L_1} \\[6pt] \dfrac{1}{L_2} \\[6pt] 0 \\[6pt] 0 \end{pmatrix} \tag{13}$$

$$A_0 = \begin{pmatrix} -\dfrac{r_1}{L_1} & 0 & -\dfrac{1}{L_1} & 0 \\[6pt] 0 & -\dfrac{r_2}{L_2} & 0 & 0 \\[6pt] \dfrac{1}{C_1} & 0 & -\dfrac{1}{RC_1} & \dfrac{1}{RC_1} \\[6pt] 0 & 0 & \dfrac{1}{RC_2} & -\dfrac{1}{RC_2} \end{pmatrix}, \quad B_0 = \begin{pmatrix} \dfrac{1}{L_1} \\[6pt] \dfrac{1}{L_2} \\[6pt] 0 \\[6pt] 0 \end{pmatrix} \tag{14}$$

The output of the voltage PI voltage controller providing the current reference for the control signal $r_s(i_1 - i_2)$ can be expressed as follows

$$r_s i_{\text{ref}} = k_p(v_{\text{ref}} - C^\mathsf{T} x) + W_i v_i, \tag{15}$$

Therefore, the switching condition when the signal $r_s(i_1 - i_2)$ reaches its peak value $r_s i_{\text{ref}} - v_{\text{ramp}}$ within a switching cycle is given by

$$k_p(v_{\text{ref}} - C^\mathsf{T} x) + W_i v_i - (v_{\text{ramp}}(t) + r_s(i_1 - i_2)) = 0, \tag{16}$$

which can be expressed in the following form

$$k_p v_{\text{ref}} + K x(t) + W_i v_i(t) - v_{\text{ramp}}(t) = 0, \tag{17}$$

where $K = (-r_s \quad r_s \quad -k_p \quad k_p)$ is the vector of feedback coefficients.

5. Accurate Stability Analysis Using Floquet Theory

The differential equations describing the dynamics of switching converters are time periodic with the switching period T determining the periodicity of solutions at the fast switching scale. DC-AC inverters are also time periodic with the switching period T and the voltage reference period $T_g = 1/f_g$. For such time periodic systems Floquet theory can be used to study the stability of periodic orbits [53]. Here, this theory will be applied using a quasi-static approximation treating the DC-AC inverter as a DC-DC converter with a slowly varying reference voltage and duty cycle. With this approximation, the reference voltage v_{ref} is considered constant within a switching cycle.

Floquet theory has been widely used in the analysis of stability of dynamical systems [53] in general and switching converters in particular [38–40]. For DC-DC converters, the stability dynamics at the fast switching cycle can be accurately predicted by analyzing the stability of the fixed points of the Poincaré map of the system using its Jacobian matrix or using Floquet theory combined with Filippov method which leads to the same results as the Poincaré map [38]. The main tool for studying the stability of periodic orbits using Floquet theory is the principal fundamental matrix or the monodromy matrix $\mathbf{M}$. This matrix plays a key role in the accurate stability analysis of switching systems [38–40,53]. The monodromy matrix is such that the dynamics in the vicinity of a quasi-static periodic orbit can be expressed as follows

$$\hat{\mathbf{x}}(t+T) = \mathbf{M}\hat{\mathbf{x}}(t) \qquad \forall t \tag{18}$$

where the overhat stands for small signal variations. Its eigenvalues are called the *characteristic multipliers* or *Floquet multipliers* and it can be seen that they determine the amount of contraction or expansion near a periodic orbit and hence they determine the stability of these periodic orbits.

Let us start by finding the monodromy matrix $\mathbf{M}$. Let $\mathbf{x}(t) \approx \mathbf{x}(t+T)$ the quasi-steady-state value of the state vector. Let $\mathbf{x}(DT) = (\mathbf{I} - \mathbf{\Phi})^{-1}\mathbf{\Psi} \approx \mathbf{x}(t) \approx \mathbf{x}(DT)$ be the value of $\mathbf{x}(t)$ at time instant DT, where $\mathbf{\Phi} = \mathbf{\Phi}_1\mathbf{\Phi}_0$, $\mathbf{\Phi}_1 = e^{\mathbf{A}_1 DT}$, $\mathbf{\Phi}_0 = e^{\mathbf{A}_0(1-D)T}$, $\mathbf{\Psi}_1 = (e^{\mathbf{A}_1 DT} - \mathbf{I})\mathbf{A}_1^{-1}\mathbf{B}V_g$, $\mathbf{\Psi}_0 = (e^{\mathbf{A}_0(1-D)T} - \mathbf{I})\mathbf{A}_0^{-1}\mathbf{B}V_g$, $\mathbf{\Psi} = \mathbf{\Phi}_1\mathbf{\Psi}_0 + \mathbf{\Psi}_1$. Let $\mathbf{m}_1(\mathbf{x}(t)) = \mathbf{A}_1\mathbf{x}(t) + \mathbf{B}_1 V_g$ and $\mathbf{m}_0(\mathbf{x}(t)) = \mathbf{A}_0\mathbf{x}(t) + \mathbf{B}_0 V_g$ be the vector fields for $u = 1$ and $u = 0$ respectively. Let us define the augmented state vector $\mathbf{x}_a = (i_1, i_2, v_{o1}, v_{o2}, v_i)^{\mathsf{T}}$. Let $\mathbf{A}_{a1}$, $\mathbf{A}_{a0}$, $\mathbf{B}_{a1}$, $\mathbf{B}_{a0}$, $\mathbf{w}_a$ and $\mathbf{K}_a$ be, respectively, the associated augmented state matrices, input vectors, vector of external parameters and vector of feedback coefficients that are expressed as follows

$$\mathbf{A}_{a1} = \begin{pmatrix} \mathbf{A}_1 & 0 \\ -1 & 0 \end{pmatrix}, \; \mathbf{B}_{a1} = \begin{pmatrix} \mathbf{B}_1 & 0 \\ 0 & 1 \end{pmatrix} \tag{19}$$

$$\mathbf{A}_{a0} = \begin{pmatrix} \mathbf{A}_0 & 0 \\ -1 & 0 \end{pmatrix}, \; \mathbf{B}_{a0} = \begin{pmatrix} \mathbf{B}_0 & 0 \\ 0 & 1 \end{pmatrix} \tag{20}$$

$$\mathbf{K}_a = \begin{pmatrix} \mathbf{K} & W_i \end{pmatrix}, \mathbf{w}_a = \begin{pmatrix} V_g \\ v_{\text{ref}} \end{pmatrix} \tag{21}$$

Let us also define the augmented state transition matrices $\mathbf{\Phi}_{a1} = e^{\mathbf{A}_{a1} DT}$ and $\mathbf{\Phi}_{a0} = e^{\mathbf{A}_{a0}(1-D)T}$ and the augmented vector fields $\mathbf{m}_{a1}(\mathbf{x}_a(t)) = \mathbf{A}_{a1}\mathbf{x}_a(t) + \mathbf{B}_{a1}\mathbf{w}_a$ and $\mathbf{m}_{a0}(\mathbf{x}_a(t)) = \mathbf{A}_{a0}\mathbf{x}_a(t) + \mathbf{B}_{a0}\mathbf{w}_a$. Then, the full-order monodromy matrix can be expressed as follows [38]

$$\mathbf{M} = \mathbf{\Phi}_{a0}\mathbf{S}\mathbf{\Phi}_{a1}, \tag{22}$$

where $\mathbf{S}$ is the saltation matrix adapted from [38] as follows

$$\mathbf{S} = \mathbf{I} + \frac{(\mathbf{m}_{a0}(\mathbf{x}_a(DT)) - \mathbf{m}_{a1}(\mathbf{x}_a(DT)))\mathbf{K}_a^{\mathsf{T}}}{W_i(v_{\text{ref}} - v_o(DT)) + \mathbf{K}^{\mathsf{T}}\mathbf{m}_1(\mathbf{x}(DT)) - m_{\text{ramp}} - m_{\text{ref}}}. \tag{23}$$

where $m_{\text{ramp}} = V_M/T$ is the slope of the ramp compensator and $m_{\text{ref}} = k_p V_{\text{ref}} 2\pi f_g \cos(2\pi f_g DT)$ is the slope contributed by the time variation of the sinusoidal voltage reference. The expression of $v_i(DT)$, the third component of $\mathbf{x}_a(DT)$, can be obtained from (17) in steady-state which gives the following expression for $v_i(DT)$

$$v_i(DT) = \frac{1}{W_i}\left(m_{\mathrm{ramp}}DT - \mathbf{K}^\mathsf{T}\mathbf{x}(DT) - k_p v_{\mathrm{ref}}\right) \tag{24}$$

Now that the expression of the monodromy matrix was derived, hereinafter, we will pay special attention to the movement of the Floquet multipliers as the voltage reference v_{ref} varies quasi-statically. This is equivalent to changing the phase angle φ or the quasi-steady-state duty cycle D. We will also study the movement of the Floquet multipliers when the proportional gain k_p of the controller or the amplitude of the ramp compensator V_M are varied. Any crossing from the interior of the unit circle to its exterior indicates a lost of stability of the desired orbit. The system becomes unstable, if at least one root of the Floquet multiplier leaves the unit circle, which is equivalent to an eigenvalue $\mathbf{M}$ leaving the unit circle. Thus, for the stability boundary $|\lambda| = 1$ for at least one eigenvalue of $\mathbf{M}$ holds. In particular, if a real characteristic multiplier goes through -1 as it moves out of the unit circle, SO at the fast switching scale takes place.

To locate the boundary of SO, the Floquet multipliers are obtained. By varying the quasi-steady-state duty cycle D, the operating point $\mathbf{x}(DT)$ was first calculated and the monodromy matrix was obtained for two different values of the proportional gain k_p. At a point where a subharmonic regime emerges, one of the eigenvalues is equal to -1. Figure 6 shows the Floquet multipliers loci in the complex plane when the quasi-steady-state duty cycle varies. The duty cycle D was varied by varying the voltage reference between 0 and its maximum values giving rise to $D \in (0.5, 0.68)$. As it can be observed from Figure 6a, for $k_p = 0.2$ all the eigenvalues remain inside the unit circle for the full considered range of the duty cycle. Then, the gain k_p was fixed at $k_p = 0.24$ then the reference voltage was varied in the same range as before and the results are depicted in Figure 6b. It can be observed that one of the eigenvalues of the monodromy matrix crosses the unit circle from the point $(-1, 0)$ in the complex plane indicating SO at a certain value of v_{ref} very close to its maximum value. The critical value of k_p at which this starts taking place is $k_p \approx 0.22$ which is in a remarkable agreement with the time-domain numerical simulations presented in Section 3.

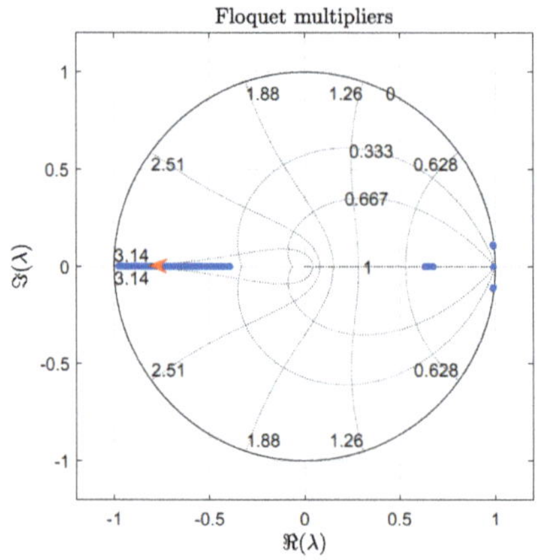

(**a**) $V_{\mathrm{ref}} \in (0, 230\sqrt{2})$ V ($D \in (0.5, 0.68)$) and $k_p = 0.20$.

Figure 6. *Cont.*

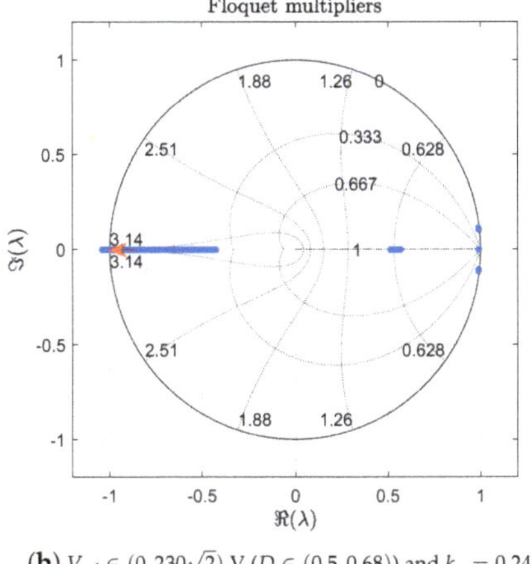

(b) $V_{\text{ref}} \in (0, 230\sqrt{2})$ V $(D \in (0.5, 0.68))$ and $k_p = 0.24$.

Figure 6. Floquet multipliers loci by varying the quasi-steady-state duty cycle D for two different values of the proportional gain k_p.

6. Stability Boundaries in the Parameter Space

If SO boundary is of concern, the expression of the characteristic equation $\det(\mathbf{M} - \lambda\mathbf{I}) = 0$ can be used by imposing that an eigenvalue $\lambda = -1$ and solving the resulting equation in a suitable projection of the parametric space. Therefore, to determine the boundary of SO, the following equation is solved for a certain system parameter after fixing the other ones

$$\det(\mathbf{M} + \mathbf{I}) = 0 \tag{25}$$

The great advantage of using (25) is that only this equation has to be solved without the need of computing all eigenvalues of $\mathbf{M}$ explicitly. Therefore, instead of solving for all eigenvalues of $\mathbf{M}$, only (25) is solved, hence, the saving of computational load is significant when the stability boundary is to be determined.

Figure 7 shows the stability boundary resulted from solving (25) with respect to the proportional gain k_p for values of the duty cycle within the operating range $(0.5, 0.68)$ and for a value of the ramp compensator amplitude $V_M = 2$ V. Within one sinewave signal one has 2000 switching cycles. Therefore, the plot was generated using 1000 values of the duty cycles and the critical values of k_p were registered in terms of D. In particular, for $V_M = 2$ V, the critical value of the proportional gain guaranteeing that all the eigenvalues lie inside the unit circle for all values of the operating duty cycle is about 0.2. This is in perfect agreement with the numerical simulations presented in Section 3. If V_M is increased, the critical value of the proportional gain also increases and the stability region gets wider as depicted in Figure 8. In particular, for $V_M = 3$ V, the critical value of the proportional gain is about 0.73, for $V_M = 4$ V, is about 1.28 and for $V_M = 5$ V, it is about 1.82. Notice that for a fixed switching period T, changing the ramp amplitude is equivalent to changing its slope.

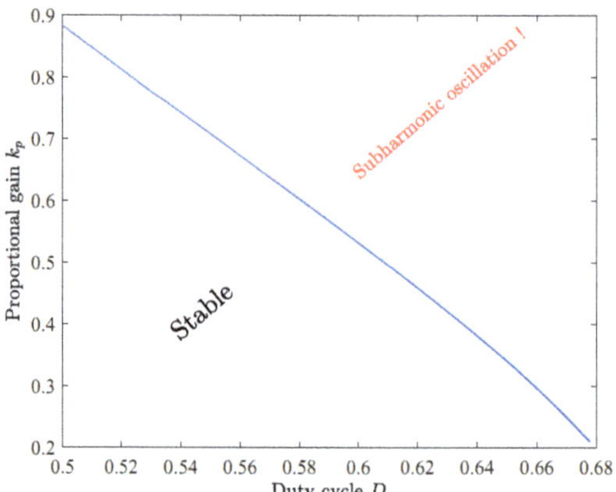

Figure 7. Stability boundaries in terms of the proportional gain k_p and the quasi-steady-state duty cycle D and for $V_M = 2$ V.

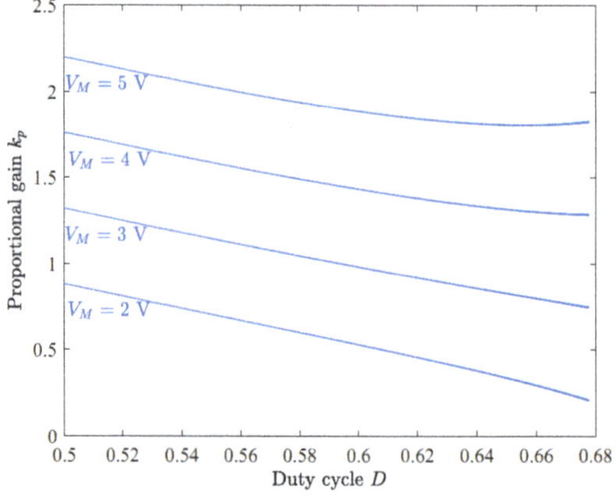

Figure 8. Stability boundaries in terms of the proportional gain k_p and the quasi-steady-state duty cycle D for different values of the ramp amplitude V_M.

As stated previously, in DC-AC inverters, the reference voltage is a time varying sinusoidal signal and accordingly the steady-state quasi-static duty cycle D is given by (8). In such a situation, the phase φ is a quasi-static parameter like D. Solving (9) in terms of the phase angle φ, one gets two critical values of the phase angle that can be expressed as follows

$$\varphi_1 \;=\; \sin^{-1}\left(\frac{V_g(2D-1)}{V_{\text{ref}}D(1-D)}\right) \tag{26}$$

$$\varphi_2 \;=\; \pi - \sin^{-1}\left(\frac{V_g(2D-1)}{V_{\text{ref}}D(1-D)}\right) \tag{27}$$

These closed expressions for the critical phase angles at which SO develops explain the observation made in Section 3. In terms of the inverter gain $M(D)$ and its maximum value M_{max}, the expressions of the phase angles are given by

$$\varphi_1 = \sin^{-1}\left(\frac{M(D)}{M_{\mathrm{max}}}\right) \tag{28}$$

$$\varphi_2 = \pi - \sin^{-1}\left(\frac{M(D)}{M_{\mathrm{max}}}\right) \tag{29}$$

The stability boundary of the system is plotted in Figure 9, in terms of the proportional gain k_p of the voltage controller and the phase angle $\varphi \in (0, \pi)$. Vertical dashed lines in Figure 9 indicate this theoretical critical value for the set of parameter values shown in Table 1. For each specific union of φ_1 and φ_2 curves, it can be noted that there is a turning point at the left side of the union. The system will be stable at the left of the turning point and will exhibit an SO phenomenon at its right side. For instance, let $k_p = 0.2$; the system is stable during the entire sinewave cycle as already observed in Figure 5a. When the proportional gain k_p is increased beyond its critical value, SO takes place within a certain phase interval, the length of which is determined by the intersection points between vertical lines corresponding to specific values of k_p and the two curves of φ_1 and φ_2. Notice that the length of the SO interval gets larger when the proportional gain increases. For instance, for $k_p = 0.4$, it is expected from Figure 9 that the system will exhibit SO in the phase interval $(\varphi_1, \varphi_2) = (46°, 134°)$ which is in close agreement with the numerical simulation depicted in Figure 5b. For $k_p = 0.6$, the expected SO interval is $(\varphi_1, \varphi_2) = (24°, 156°)$ which is in close agreement with Figure 5c and for $k_p = 0.8$, the expected SO interval is $(\varphi_1, \varphi_2) = (7°, 173°)$ which is in close agreement with Figure 5d.

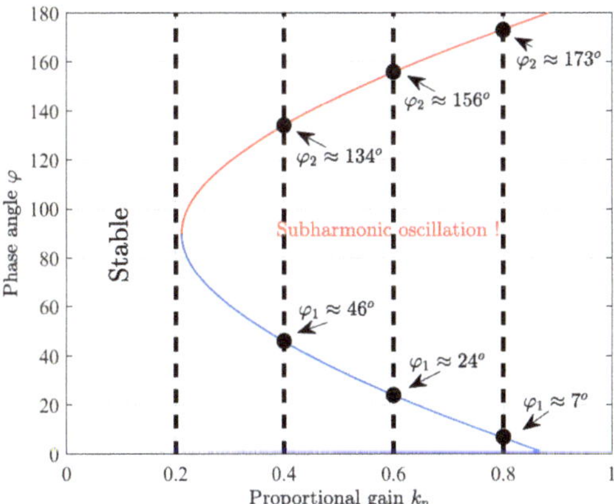

Figure 9. Critical phase angles in [°] defining the SO interval in terms of the proportional gain k_p.

The estimated values of the critical phase angles from Figure 9 defining the SO interval differ slightly from the numerical simulation result in Figure 5. The discrepancies between the theoretically predicted values in Figure 9 and the ones obtained from numerical simulations depicted in Figure 5 can be attributed to two main factors. The first one is the use of the quasi-static approximation. The second one is the fact that at the point where bubbling develops its amplitude is extremely small making it invisible in the scale used for representing the complete waveforms of the duty cycle during one entire sinewave cycle. By zooming close the critical values of φ, more accurate data can be obtained and discrepancies decrease significantly.

As has been shown in Figure 8, the maximal value of the proportional gain k_p guaranteeing stability during the entire the sinewave cycle depends on the ramp amplitude V_M. Therefore, the critical phase angle curves depicted in Figure 9 are also obtained for different values of V_M and the results are depicted in Figure 10. For each ramp amplitude V_M, a value of the proportional gain k_p selected at the left of the corresponding turning point will guarantee no presence of SO during the entire sinewave cycle. Note that as the ramp amplitude V_M increases the maximal value allowed for the proportional gain k_p also increases.

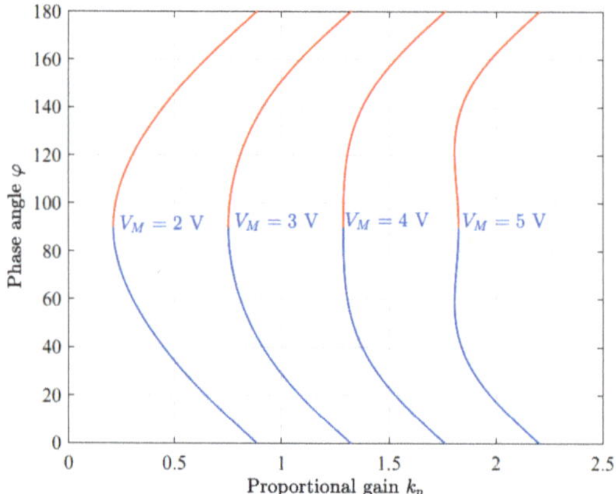

Figure 10. Critical phase angles in [°] defining the SO interval in terms of the proportional gain k_p for different values of the ramp amplitude V_M.

7. Conclusions

This paper has focused on the subharmonic oscillation boundary leading to bubbling phenomenon in a single-phase DC-AC differential boost inverter with a linear resistive load. This work has provided a comprehensive study of the system and stability problems of the system were discussed in order to determine stabilizing parameter space. This facilitates convenient selection of parameter values to avoid distortion due to subharmonic oscillation instability in some intervals of the sinewave voltage reference. Therefore, the results are useful for practical design of DC-AC inverters to ensure a stable operation and hence maintain a high power quality and ensuring low and acceptable values of THD. By using time domain waveforms computed from the circuit-level switched model of the system, it was shown that the differential boost inverter could exhibit subharmonic oscillation instabilities at the fast switching scale. Stable and unstable zones of operation, critical parameter values and stability boundaries have been determined. Floquet theory combined with quasi-static approximation has been used resulting in accurately locating the critical values of the system parameters. The theoretical predictions are in perfect agreement with the results obtained from numerical simulations performed on the circuit-level switched model of the inverter. The methodology presented in this study can be applied to other inverter topologies.

Author Contributions: Conceptualization, A.E.A.; methodology, A.E.A. software, A.E.A.; validation, M.A.-N.; formal analysis, A.E.A.; investigation, A.E.A.; resources, M.A.-N.; data curation, R.H.; writing—original draft preparation, A.E.A.; writing—review and editing, M.A.-N. and M.H.; visualization, M.A.-N., R.H. and M.H.; supervision, M.A.-N.; funding acquisition, M.A.-N. All authors have read and agreed to the published version of the manuscript.

Funding: This work has been sponsored by the Spanish Agencia Estatal de Investigación (AEI) and the Fondo Europeo de Desarrollo Regional (FEDER) under grants DPI2017-84572-C2-1-R. A. El Aroudi and M. Al-Numay acknowledge financial support from the Researchers Supporting Project number (RSP-2020/150), King Saud University, Riyadh, Saudi Arabia.

Conflicts of Interest: The authors declare no conflict of interest.

Abbreviations

The following abbreviations are used in this manuscript:

AC	Alternate Current
DC	Direct Current
CMC	Current Mode Control
PWM	Pulse Width Modulation
SO	Subharmonic Oscillation

References

1. Yang, C.-Y.; Hsieh, C.-Y.; Feng, F.-K.; Chen, K.-H. Highly efficient analog maximum power point tracking (AMPPT) in a photovoltaic system. *IEEE Trans. Circuits Syst. Reg. Pap.* **2012**, *59*, 1546–1556. [CrossRef]
2. Wai, R.-J.; Wang, W.-H. Grid-connected photovoltaic generation system. *IEEE Trans. Circuits Syst. Reg. Pap.* **2008**, *55*, 953–964.
3. Fang, Y.; Zhu, Y.; Fei, J. Adaptive intelligent sliding mode control of a photovoltaic Grid-connected inverter. *Appl. Sci.* **2018**, *8*, 1756. [CrossRef]
4. Sakharuk, T.A.; Stankovic, A.M.; Tadmor, G.; Eirea, G. Modeling of PWM inverter-supplied AC drives at low switching frequencies. *IEEE Trans. Circuits Syst. Fundam. Theory Appl.* **2002**, *49*, 621–631. [CrossRef]
5. Lee, K.; Ha, J.-I. Single-phase inverter drive for interior permanent magnet machines. *IEEE Trans. Power Electron.* **2017**, *32*, 1355–1366 . [CrossRef]
6. Gupta, R.; Ghosh, A. Frequency-domain characterization of sliding mode control of an inverter used in DSTATCOM application. *IEEE Trans. Circuits Syst. Reg. Pap.* **2006**, *53*, 662–676. [CrossRef]
7. Kawamura, A.; Chuarayapratip, R.; Haneyoshi, T. Deadbeat control of PWM inverter with modified pulse patterns for uninterruptible power supply. *IEEE Trans. Ind. Electron.* **1988**, *35*, 295–300. [CrossRef]
8. Bandyopadhyay, A.; Howrah, S.; Mandal, K. Design-oriented dynamical analysis of single-phase H-bridge inverter. In Proceedings of the 2020 IEEE International Conference on Power Electronics, Smart Grid and Renewable Energy (PESGRE2020), Cochin, India, 2–4 January 2020.
9. El Aroudi, A.; Al-Numay, M.S.; Lu, W.G.; Bosque-Moncusi, J.M.; Iu, H.H. A combined analytical-numerical methodology for predicting subharmonic oscillation in H-bridge inverters under double edge modulation. *IEEE Trans. Circuits Syst. Reg. Pap.* **2018**, *65*, 2341–2351. [CrossRef]
10. Avrutin, V.; Mosekilde, E.; Zhusubaliyev, Z.T.; Gardini, L. Onset of chaos in a single-phase power electronic inverter. *Chaos Interdisc. J. Nonlinear Sci.* **2015**, *25*, 43114. [CrossRef]
11. El Aroudi A.; Orabi A.; Martinez-Salamero, L. A representative discrete-time model for uncovering slow and fast scale instabilities in boost power factor correction AC-DC pre-regulators. *Int. J. Bifurc. Chaos* **2008**, *18*, 3073–3092. [CrossRef]
12. Shankar, D.P.; Govindarajan, U.; Karunakaran, K. Period-bubbling and mode-locking instabilities in a full-bridge dc-ac buck inverter. *IET Power Electron.* **2013**, *6*, 1956–1970. [CrossRef]
13. Robert, B.; Robert, C. Border collision bifurcations in a one-dimensional piecewise smooth map for a PWM current-programmed h-bridge inverter. *Int. J. Control.* **2002**, *75*, 1356–1367. [CrossRef]

14. Asahara, H.; Kousaka, T. Bifurcation analysis in a PWM current controlled H-bridge inverter. *Int. J. Bifurc. Chaos* **2011**, *21*, 985–996. [CrossRef]

15. Li, M.; Dai, D.; Ma, X. Slow-scale and fast-scale instabilities in voltage-mode controlled full-bridge inverter. *Circuits Syst. Signal Process.* **2008**, *27*, 811–831. [CrossRef]

16. El Aroudi, A.; Rodriguez, E.; Orabi, M.; Alarcon, E. Modeling of switching frequency instabilities in buck-based dc-ac H-bridge inverters. *Int. J. Circuit Theory Appl.* **2011**, *39*, 175–193. [CrossRef]

17. Zhusubaliyev, Z.T.; Mosekilde, E.; Andriyanov, A.I.; Shein, V.V. Phase synchronized quasiperiodicity in power electronic inverter systems. *Phys. Nonlinear Phenom.* **2014**, *268*, 14–24. [CrossRef]

18. Avrutin, V.; Morcillo, J.D.; Zhusubaliyev, Z.T.; Angulo, F. Bubbling in a power electronic inverter: Onset, development and detection. *Chaos Solitons Fractals.* **2017**, *104*, 135–152. [CrossRef]

19. Lei, B.; Xiao, G.; Wu, X.; Ka, Y.R.; Zheng, L. Bifurcation analysis in a digitally controlled H-bridge grid-connected inverter. *Int. J. Bifurc. Chaos* **2014**, *24*, 1450002. [CrossRef]

20. Benadero, L.; Ponce, E.; El Aroudi, A.; Torres, F. Limit cycle bifurcations in resonant LC power inverters under zero current switching strategy. *Nonlinear Dyn.* **2018**, *91*, 1145–1161. [CrossRef]

21. Albea, C.; Gordillo, F. Estimation of the region of attraction for a boost DC-AC converter control law. In Proceedings of the IFAC Symposium on Nonlinear Control Systems (Nolcos), Pretoria, South Africa, 21–24 August 2007; pp. 874–879.

22. Liao, Z.-X.; Luo, D.; Luo, X.-S.; Li, H.-S.; Xiang, Q.-Q.; Huang, G.-X.; Li, T.-H.; Jiang, P.-Q. Nonlinear model and dynamic behavior of photovoltaic grid-connected inverter. *Appl. Sci.* **2020**, *10*, 2120. [CrossRef]

23. Lu, W.G.; Jing, F.; Zhou, L.; Iu, H.H.C.; Fernando, T. Control of sub-harmonic oscillation in peak current mode buck converter with dynamic resonant perturbation. *Int. J. Circuit Theory Appl.* **2015**, *43*, 1399–1411. [CrossRef]

24. El Aroudi, A.; Mandal, K.; Giaouris, D.; Banerjee, S.; Abusorrah, A.; Al Hindawi, M.; Al-Turki, Y. Fast-scale stability limits of a two-stage boost power converter. *Int. J. Circuit Theory Appl.* **2016**, *44*, 1127–1141. [CrossRef]

25. El Aroudi, A. A new approach for accurate prediction of subharmonic oscillation in switching regulators–part I: Mathematical derivations. *IEEE Trans. Power Electron.* **2017**, *32*, 5651–5665. [CrossRef]

26. El Aroudi, A. A new approach for accurate prediction of subharmonic oscillation in switching regulators–part II: Case studies. *IEEE Trans. Power Electron.* **2017**, *32*, 5835–5849. [CrossRef]

27. El Aroudi, A.; Al-Numay, M.; Calvente, J.; Giral, R.; Rodriguez, E.; Alarcón, E. Prediction of subharmonic oscillation in switching regulators: from a slope to a ripple standpoint. *Int. J. Electron.* **2016**, *103*, 2090–2109. [CrossRef]

28. Huang, M.; Ji, H.; Sun, J.; Wei, L.; Zha, X. Bifurcation-based stability analysis of photovoltaic-battery hybrid power system. *IEEE J. Emerg. Sel. Top. Power Electron.* **2017**, *5*, 1055–1067. [CrossRef]

29. Huang, M.; Peng, Y.; Tse, C.K.; Liu, Y.; Sun, J.; Zha, X. Bifurcation and large-signal stability analysis of three-phase voltage source converter under grid voltage dips. *IEEE Trans. Power Electron.* **2017**, *32*, 8868–8879. [CrossRef]

30. Peng, D.; Huang, M.; Li, J.; Sun, J.; Zha, X.; Wang, C. Large-signal stability criterion for parallel-connected DC–DC converters with current source equivalence. *IEEE Trans. Circuits Syst. II Express Briefs* **2019**, *66*, 2037–2041. [CrossRef]

31. Cheng, L.; Ki, W.-H.; Yang, F.; Mok, P.K.T.; Jing, X. Predicting subharmonic oscillation of voltage-mode switching converters using a circuit-oriented geometrical approach. *IEEE Trans. Circuits Syst. Reg. Pap.* **2017**, *64*, 717–730. [CrossRef]

32. Leng, M.; Zhou, G.; Zhou, S.; Zhang, K.; Xu, S. Stability analysis for peak current-mode controlled buck LED driver Based on discrete-time modeling. *IEEE J. Emerg. Sel. Top. Power Electron.* **2018**, *6*, 1567–1580. [CrossRef]

33. Zhou, S.; Zhou, G.; Zeng, S.; Xu, S.; Ma, H. Unified discrete-mapping model and dynamical behavior analysis of current-mode controlled single-inductor dual-output DC-DC Converter. *IEEE J. Emerg. Sel. Top. Power Electron.* **2019**, *7*, 366–380. [CrossRef]

34. Banerjee, S.; Verghese, G.C. *Nonlinear Phenomena in Power Electronics Attractors, Bifurcations, Chaos, and Nonlinear Control*; IEEE Press: New York, NY, USA, 2001.

35. Tse, C.K. *Complex Behavior of Switching Power Converters*; CRC Press: New York, NY, USA, 2003.

36. Mazumder, S.K.; Nayfeh, A.H.; Boroyevich, D. Theoretical and experimental investigation of the fast- and slow-scale instabilities of a DC-DC converter. *IEEE Trans. Power Electron.* **2001**, *16*, 201–216. [CrossRef]
37. Fossas, E.; Olivar, G. Study of chaos in the buck converter. *IEEE Trans. Circuits Syst. Fundam. Theory Appl.* **1996**, *43*, 13–25. [CrossRef]
38. Giaouris, D.; Maity, S.; Banerjee, S.; Pickert, V.; Zahawi, B. Application of Filippov method for the analysis of subharmonic instability in DC-DC converters. *Int. J. Circuit Theory Appl.* **2009**, *37*, 899–919. [CrossRef]
39. Cortes, J.; Svikovic, V.; Alou, P.; Oliver, J.A.; Cobos, J.A.; Wisniewski, R. Accurate analysis of subharmonic oscillations of V^2 and $V^2 I_c$ controls applied to buck converter. *IEEE Trans. Power Electron.* **2015**, *30*, 1005–1018. [CrossRef]
40. El Aroudi, A.; Al-Numay, M.; Garcia, G.; Al Hossani, K.; Al Sayari, N.; Cid-Pastor, A. Analysis of nonlinear dynamics of a quadratic boost converter used for maximum power point tracking in a grid-interlinked PV system. *Energies* **2019**, *12*, 61. [CrossRef]
41. Jang, M.; Agelidis, V.G. A minimum power-processing-stage fuel-cell energy system based on a boost-inverter with a bidirectional backup battery storage. *IEEE Trans. Power Electron.* **2011**, *26*, 1568–1577. [CrossRef]
42. Zhu, G.; Tan, S.; Chen, Y.; Tse, C.K. Mitigation of low-frequency current ripple in fuel-cell inverter systems through waveform control. *IEEE Trans. Power Electron.* **2013**, *28*, 779–792. [CrossRef]
43. Jha, K.; Mishra, S.; Joshi, A. High-quality sine wave generation using a differential boost inverter at higher operating frequency. *IEEE Trans. Ind. Appl.* **2015**, *51*, 373–384. [CrossRef]
44. Flores-Bahamonde, F.; Valderrama-Blavi, H.; Bosque-Moncusi, J.M.; Garcia, G.; Martinez-Salamero, L. Using the sliding-mode control approach for analysis and design of the boost inverter. *IET Power Electron.* **2016**, *9*, 1625–1634. [CrossRef]
45. Lopez-Caiza, D.; Flores-Bahamonde, F.; Kouro, S.; Santana, V.; Müller, V.; Chub, A. Sliding mode based control of dual boost inverter forg grid connection. *Energies* **2019**, *12*, 4241. [CrossRef]
46. Meneses, D.; Blaabjerg, F.; García, O.; Cobos, J.A. Review and comparison of step-up transformerless topologies for photovoltaic AC-module application. *IEEE Trans. Power Electron.* **2013**, *28*, 2649–2663. [CrossRef]
47. Caceres, R.O.; Barbi, I. A boost DC–AC converter: Analysis, design, and experimentation. *IEEE Trans. Power Electron.* **1999**, *14*, 134–141. [CrossRef]
48. Sanchis, P.; Ursaea, A.; Gubia, E. Boost DC–AC inverter: a new control strategy. *IEEE Trans. Power Electron.* **2005**, *20*, 343–353. [CrossRef]
49. El Aroudi, A; Haroun, R.; Al-Numay, M.; Huang, M. Multiple-loop control design for a single-stage PV-fed grid-tied differential boost inverter. *Appl. Sci.* **2020**, *10*, 4808. [CrossRef]
50. Vazquez, N.; Almazan, J.; Alvarez, J. Analysis and experimental study of the buck, boost and buck-boost inverters. In Proceedings of the 30th Annual IEEE Power Electronics Specialists Conference. Record. (Cat. No. 99CH36321), Charleston, SC, USA, 1 July 1999; pp. 801–806.
51. Iu, H.H.C.; Zhou, Y.; Tse, C.K. Fast-scale instability in a boost PFC converter under average current control. *Int. J. Circuit Theory Appl.* **2003**, *31*, 611–624. [CrossRef]
52. Floquet, G. Sur les équations différentielles linéaires à coefficients périodiques. *Ann. Sci. L'éCole Norm. SupéRieure* **1883**, *12*, 47–88. [CrossRef]
53. Leine, R.L.; Nijemeijer, H. Dynamics and bifurcations of non-smooth mechanical systems. In *Lecture Notes in Applied and Computational Mechanics*; Springer: Berlin/Heidelberg, Germany, 2004; Volume 18.

Article

Double Sliding-Surface Multiloop Control Reducing Semiconductor Voltage Stress on the Boost Inverter

Oswaldo López-Santos [1,*] and Germain García [2]

[1]   Facultad de Ingeniería, Universidad de Ibagué, Carrera 22 Calle 69 Barrio Ambalá, 730001 Ibagué, Colombia
[2]   LAAS-CNRS, Université de Toulouse, CNRS, INSA, 31077 Toulouse, France; garcia@laas.fr
*   Correspondence: oswaldo.lopez@unibague.edu.co; Tel.: +57-8-2760010 (ext. 4007)

Received: 18 May 2020; Accepted: 13 July 2020; Published: 17 July 2020

Abstract: Sliding-mode control (SMC) has been successfully applied to boost inverters, which solves the tracking problem of imposing sinusoidal behavior to the output voltage despite the coupled or decoupled operation of both boost cells in the converter. Most of the results reported in the literature were obtained using the conventional cascade-control structure involving outer loops that generate references for one or two sliding surfaces defined using linear combinations of inductor currents and capacitor voltages. As expected, all proposed methods share the inherent robustness and insensitivity to the uncertainties of SMC, which are the reasons why one of the few comparison criteria between them is the simplicity of their implementation that is evaluated according to the required measurements and mathematical operations. Furthermore, the slight differences between the obtained dynamic performances do not allow a clear distinction of the best solution. This study presents a new SMC approach applied to a boost inverter in which two boost cells are independently commutated. Each of these boost cells integrates an outer loop, enforcing the tracking of harmonic-enriched waveforms to the capacitor voltage. Although this approach increases by two the number of measurements and requires multiloop controllers, it allows effective alleviation of the semiconductor voltage stress by reducing the required voltage gain. A complete analytical study using harmonic balance technique allows deducing a simplified model allowing to obtain a PI controller valid into to the whole set of operation conditions. The several simulation results completely verified the potential of the control proposal and the accuracy of the employed methods.

Keywords: boost inverter; harmonic balance; sliding mode control

1. Introduction

The DC–AC boost converter, boost inverter, differential boost inverter or dual boost inverter, as it has been called by different authors in the literature, was introduced by Caceres and Barbi in the 1990s [1]. The potential that allowed this converter to attract the interest of researchers in the following decades was its ability to overcome the most restrictive limitations of the conventional and well-established full-bridge inverter. This is mainly related to the possibility of generating AC voltages with amplitudes larger than the input DC voltage without requiring more than one conversion stage or increasing the number of power semiconductors. This interesting topology is composed of two identical cells using symmetrical bidirectional boost DC–DC converters that share a back-to-back or bridge connection and differentially provide the output voltage between their outputs. This feature is characterized by a reduced common-mode noise but can be affected by DC current circulation in the load if the bias component of the output voltage of the cells is not correctly equalized [2].

Although the aforementioned features of the boost inverter are very attractive, these can only be achieved by developing more complex controllers or applying more complex methods to synthesize the required controllers. Compared with its counterpart, i.e., the full-bridge inverter, the highly nonlinear dynamic behavior of the output voltage as a function of the operating duty cycle makes devising an effective control solution difficult. The higher the required gain to generate the output voltage is, the higher is the effect of nonlinearity on the dynamic behavior causing that linear controllers cannot guarantee a proper operation. Further, we need to mention that the approach performance based on linear control is directly affected by the non-minimum phase nature of the output voltage to control the transfer function of each boost cell.

Sliding-mode control (SMC) was practically the first approach in the reported literature applied to control a boost converter in the context of a stand-alone operation. In [1], Caceres and Barbi defined a sliding surface for each boost cell using a linear combination of current and voltage errors. Although the required voltage references were nothing more than two DC-biased sinusoidal waveforms with a 180° phase shift between them, the current references were difficult to synthesize because it depended on the input voltage and load current. Hence, the current-error component of the surfaces was obtained by measuring the high-frequency component of the inductor currents by assuming perfect tracking of the references in the stationary state. This approach was implemented using hysteresis comparators that then generated a variable switching frequency. This first proposal could be classified into a set of double-surface SMC (DS-SMC) approaches. A second method that employed SMC was introduced by Cortes et al. in [3] in which the control of both boost cells were correlated by complementarily coupling the commutation of the switches in the boost cells. In that work, a unique sliding surface was defined that involved the difference between the inductor currents and the proportional and integral actions that both operated under an output–voltage error. This second proposal, which could be classified as a single-surface SMC, was implemented using a single hysteresis comparator and required one less measurement because the voltage of the capacitors were not separately controlled. Furthermore, in the selected sliding surface, identifying the form of the well-known indirect control applied to regulate the output voltage of DC–DC converters was possible using an outer proportional–integral (PI) controller and a current inner loop. More recently, Flores-Bahamonde et al. have reemployed the same technique by preserving the coupled control action between the converter cells and applying an equivalent control technique to provide an analytical solution for synthesizing the outer voltage controller [4]. From the perspective of that work, the converter was controlled by imposing a periodic reference to the difference between the input currents of the cells, which in turn was generated by a PI outer loop that was configured to enforce the desired shape on the output voltage. That work also employs a hysteresis comparator for implementation. Finally, from a different perspective of the SMC application in this field and using a constant-frequency modulator, Wai et al. developed an adaptive fuzzy-neural-network control (AFNNC) in combination with a total sliding-mode controller [5]. To design the controller, an additional loop, called as curbing controller, allowed modification of the sliding surface to cope with unpredictable disturbances, whereas the AFNNC modified the sliding-surface parameters to ensure permanent stability. Naturally, despite the achieved good performance, the required implementation was considerably complex compared with that in [4]. In addition, SMC has been applied to the control of boost inverters in grid-connected applications, which is similar to the work reported in [6] in which a boost-inverter-based hybrid energy-storage system (HESS) integrates both batteries and supercapacitors to the grid using two sliding-mode controllers. A particular aspect to consider regarding this work is the fact that the system can be considered as two boost inverters that share the connection to the cell capacitors and then share the AC-side differential connection. The complexity of the control in this approach then comes from the power control used to inject apparent power into the grid (both active and reactive power), the control used to impose the charge–discharge regimes of the storage devices and the compact configuration of the HESS.

More recently, following the interest in systems involving batteries and fuel cells, researchers have paid more attention to the input–current behavior of the boost inverter mainly because of the scientific results that demonstrate that the ripple content exerts an important negative effect on the lifetime of storage devices. In a stand-alone case, to adequately shape the output voltage, the control must minimize the ripple content of the input current, which constitutes an important control objective.

As expected, the simultaneous accomplishment of these two functions results in new stationary behavior in the inner variables, which is reflected in their harmonic content. In [7], Jha et al. presented a cascade-control approach based on the use of a linearization function that evolved depending on the load and input–voltage operating conditions. A particular feature of this proposal was the use of three voltage measurements that shared the same reference and that no current measurement was required. Similarly, Zhu el al. proposed in [8] a waveform control using a cascade controller with two loops: one that shaped the voltage of the capacitors of the boost cells and the other that shaped the overall output voltage. These two controllers exhibited the characteristic of avoiding the use of current measurements.

In the grid-connected case, the control of the inverter forces the shape of not only the output voltage, but also the current to inject the generated power into the grid. This application of the boost inverter was explored by Angelidis and Vassilos et al. in [9] using a cascade-control scheme that involved a loop compensation of the nonlinear gain of each boost cell and then provided the reference of the inner loops that affected the inductor currents. Similar to the stand-alone case, the enforcement of the input current to be simultaneously constant with the primary control objective implies an adjustment of the stationary behavior of the converter variables. A rule-based cascade controller that determined a component to be added to the voltage references in order to cancel the second-order harmonic of the input current was developed in [10]. The proposed scheme required the measurement of the ripple component of the input current to compute the amplitude and phase of the voltage references of the cascade loops that included a voltage outer loop and a current inner loop per boost cell. The innovative feature of this controller was the use of a perturb and observe algorithm that computed the amplitude and phase of the required additional components. Again, a similar cascade-control scheme was proposed in [11] where an additional outer loop generated the added components from the measurement of the AC component of the input current. The proposed loop separately determined the AC components to avoid undesired phase shifts. The common feature of these last three controllers was the use of harmonic-enriched references in the capacitor-voltage control loops, which increased the ability of the control system to perform a second function in addition to the tracking of the output voltage or current. Other interesting works that improved the control of boost inverters can be found into battery-charging [12,13] and fuel-cell-based applications [14,15].

This study presents an alternative DS-SMC approach that is applied to a boost inverter in which the boost cells are independently commutated using one cascade controller per cell, thereby enforcing tracking of harmonic-enriched reference waveforms to shape the capacitor voltage. Second- and fourth-order terms are introduced to reduce the instantaneous-voltage gain of the cells, which results in the alleviation of the voltage stress of the switches throughout the period of the output voltage, as previously assessed in [16] where the same converter was modified by introducing additional power semiconductors and using half-cycle rectified voltage references. In contrast to the SMC approach developed in [4], the proposed control requires two additional voltage measurements, one additional hysteresis comparator and a complementary reference generator to produce the two required harmonic components. Although this last aspect supposes an increment in terms of cost of the solution, the three required voltage measurements share the same electric reference which suppose no need of isolation. Furthermore, since isolation is preferred for increased values of nominal power of the inverter, the cost of the additional electronics will become irrelevant. Beside this, complexity of the system is not really affected because the control philosophy is the same (hysteresis comparison and PI outer loop). The size of the electronic circuit or the computational cost increase depending on the type of implementation, but this aspect is not comparable with the considerable reduction of the semiconductor voltage stress.

The rest of the study is organized as follows: The boost converter operation and the fundamentals of the performance improvement are presented in Section 2. After that, in Section 3, the ideal sliding motion is analyzed for the entire range of operation conditions using the harmonic balance technique. The constraints to ensure asymptotic stability when tracking the required periodic behavior are provided as well as a simplified model of the inner loop dynamics. The fundamentals for synthesis and implementation of the outer controllers are also explained. Comparative simulation results demonstrate the feasibility and advantages of the proposed control in Section 4. Finally, conclusions and future work are presented in Section 5.

2. Fundamentals of the Performance Improvement

The boost inverter depicted in Figure 1 is a fourth order DC–AC converter which is composed by two conventional boost converter cells. The sub-indices $x = \{1,2\}$ were defined to differentiate left and right sides of the converter, respectively. Each cell is integrated by one inductor (L), one capacitor (C) and two controlled switches (S_{x1}: high-side and S_{x2}: low-side). The two switches into a cell assemble a bridge leg and then the complete circuit topology is quite similar to the conventional full-bridge inverter. The input voltage v_{in} is applied to the inputs of the two cells and the output voltage (v_o) is obtained as the difference between the voltages of the output capacitors of the cells (v_1 and v_2). The study is developed considering a resistive load r_o.

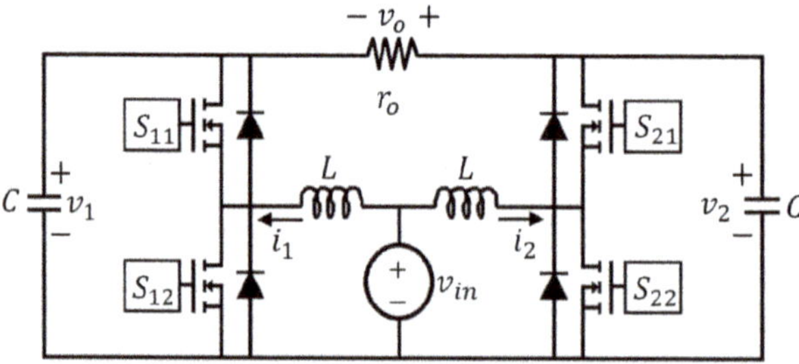

Figure 1. Schematic diagram of the boost inverter circuit.

Stand-alone applications demand that inverter produces a pure sinusoidal signal from a DC input voltage which in the case of the boost inverter can be lower that the amplitude of the desired output. Then consider the following output voltage:

$$v_{oe} = V_m \sin \omega t = 2V_{s1} \sin \omega t \tag{1}$$

being V_m the desired amplitude and $\omega = 2\pi f$, for f being the desired output frequency. To obtain this voltage, majority of control proposals define as objective the tracking of DC-biased sinusoidal references in the capacitor voltages. These references have the form:

$$v_{1e} = V_{dc} - V_{s1} \sin \omega t \tag{2}$$

$$v_{2e} = V_{dc} + V_{s1} \sin \omega t \tag{3}$$

Considering that Equation (1) is obtained from the difference between Equations (3) and (2), it is easy to note that the same result can be produced using references v_{1e} and v_{2e} with enriched harmonic content as follows:

$$v_{1e} = V_{dc} - V_{s1} \sin \omega t + V_{c1} \cos \omega t + \sum_{n=2}^{\infty} V_{sn} \sin n\omega t + V_{cn} \cos n\omega t \tag{4}$$

$$v_{2e} = V_{dc} + V_{s1} \sin \omega t + V_{s1} \cos \omega t + \sum_{n=2}^{\infty} V_{sn} \sin n\omega t + V_{cn} \cos n\omega t \tag{5}$$

As it was developed in [16], the voltage stress of the semiconductors in the boost inverter is directly related with the shape of the capacitor voltages. Then, that work shows how enforcing a DC-biased half-wave rectified sinusoidal is optimal to alleviate the switching losses improving the efficiency of the converter. However, in that work, the desired behavior is obtained by using two additional switches increasing the number of semiconductors and the required auxiliary circuitry. Then, the concept developed in this study consists of enforcing the desired shape in the capacitor voltages through independent controllers for each boost cell of the inverter simultaneously ensuring an output voltage of high quality. Then, two objectives are defined: (a) generating the adequate voltage references and (b) guaranteeing a robust tracking of these references.

Consider that references are defined to enforce the DC-biased half-wave rectified sinusoidal shape in the capacitor voltages as follows:

$$v_{1e} = V_{dc} - V_{s1} + 2V_{s1} \sin \omega t [1 + \text{sign}(\sin \omega t)] \tag{6}$$

$$v_{2e} = V_{dc} + V_{s1} + 2V_{s1} \sin \omega t [1 + \text{sign}(\sin \omega t)] \tag{7}$$

Although references Equations (6) and (7) can be easily produced, they can introduce negative effects into the control loop because of the discontinuity of their derivatives. Then, it is proposed to build approximate, but smooth references by adding one or two harmonic components to Equations (2) and (3). To sake of simplicity, analysis is shown only for reference signal v_{2e} as follows:

$$v_{2e} = V_{dc} - V_{s1} + V_{c2} + V_{s1} \sin \omega t - V_{c2} \cos 2\omega t \tag{8}$$

$$v_{2e} = V_{dc} - V_{s1} + V_{c2} + V_{c4} + V_{s1} \sin \omega t - V_{c2} \cos 2\omega t - V_{c4} \cos 4\omega t \tag{9}$$

Figure 2 depicts a comparison between signals obtained from numeric evaluation of Equations (7)–(9). Although a higher number of harmonics can improve the results, solutions given by Equations (8) and (9) are acceptably good. As it can be observed, for Equation (8) the signal remains below the pure sinusoidal by a considerable margin while the difference with respect to the signal Equation (9) is slight, but also important.

Deduction of the amplitude of the added components is constrained to obtain a reference which maximum and minimum values be the same of the pure sinusoidal. A numeric analysis allows deduce that for Equation (8), $V_{c2} = 0.256 V_{s1}$ is optimal, while for Equation (9), $V_{c2} = 0.36 V_{s1}$ and $V_{c4} = 0.036 V_{s1}$ are optimal.

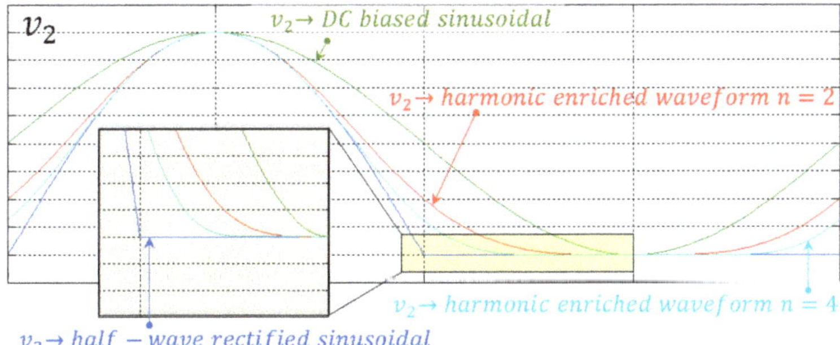

Figure 2. Comparison of possible waveforms for v_2 constraining the voltage excursion.

The minimum limits of the waveforms in the graphic comparison are the same because $V_{dc} = 0$ for Equation (7), $V_{dc} = 0.75 V_{s1}$ for Equation (8) and $V_{dc} = 0.676 V_{s1}$ for Equation (9). Consequently, the value of V_{dc} into the references must be defined adequately to ensure that this limit be always higher that the instantaneous input voltage.

3. Control of the Inverter

In this work, the adopted control scheme is a classical cascade control structure. The two inner loops, one for each boost cell, are current control loops. Due to its great flexibility and ease of implementation, a sliding mode control strategy was retained. The outer loops, one for each boost cell, are voltage control loops based on saturated PI controllers whose outputs provide an adequate current reference for the inner loop controllers. The need of saturated PI controllers will be justified later. An important step for the success of the adopted control strategy is the generation of appropriate voltage references considering the performance improvements discussed and proposed in Section 2. These voltage references are determined by a harmonic balance method applied to a power balance which constitutes a constraint imposed by the inverter structure. Figure 3 depicts the overall controlled system. Note that the sliding modes controllers are implemented by using simple hysteresis comparators leading to inner controls expressed as:

$$u_x = \begin{cases} 0 & \text{if} \quad S_x(x) > \delta \\ 1 & \text{if} \quad S_x(x) < -\delta \end{cases} \quad x = 1, 2$$

where δ is a small positive number defined to constraint the maximum switching frequency of the converter and $S_x(x)$ are the sliding surfaces introduced in a next paragraph. Note that possible variations of the input source voltage is considered in the proposed strategy to modify the DC-component of the voltage references. This aspect is covered by taking measure of $v_{in}(t)$ and using a low-pass (LP) filter to smooth the effect of ripple content. The cutoff frequency of the LP filter must be selected accordingly with the output frequency (three or four times is enough).

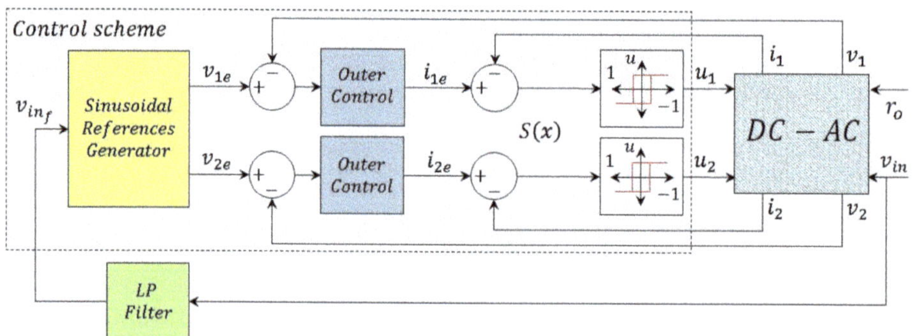

Figure 3. Block diagram of the complete control proposal: reference-generation and control.

3.1. Model of the Inverter

The model of the inverter is deduced from the four circuit structures presented in Figure 4. Introducing the control signals u_1 and u_2, corresponding to $u_x = 0$ if S_{x1} is on and S_{x2} is off and conversely $u_x = 1$ if S_{x1} is off and S_{x2} is on, $x = 1, 2$, the converter circuit can be modeled by means of the following state equations:

$$\frac{di_1(t)}{dt} = \frac{1}{L}v_{in}(t) - \frac{1-u_1}{L}v_1(t) \tag{10}$$

$$\frac{di_2(t)}{dt} = \frac{1}{L}v_{in}(t) - \frac{1-u_2}{L}v_2(t) \tag{11}$$

$$\frac{dv_1(t)}{dt} = \frac{1-u_1}{C}i_1(t) + \frac{1}{r_0C}v_o(t) \tag{12}$$

$$\frac{dv_2(t)}{dt} = \frac{1-u_2}{C}i_2(t) - \frac{1}{r_0C}v_o(t) \tag{13}$$

$$v_o(t) = v_2(t) - v_1(t) \tag{14}$$

where $\mathbf{u}_1$ is the control signal of the left-side boost cell and $\mathbf{u}_2$ is the control signal of the right-side cell.

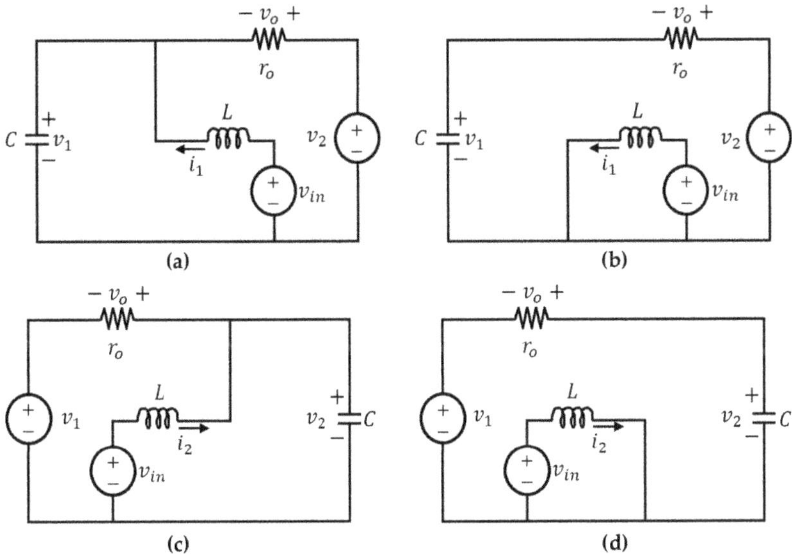

Figure 4. Circuit structures of the boost inverter operating in DS-SMC. (**a**) Cell 1 state 0, (**b**) cell 1 state 1, (**c**) cell 2 state 0 and (**d**) cell 2 state 1.

3.2. Inner Current Control Loops

To enforce a sliding mode regime, the switches of the boost inverter can be operated using two different commutation techniques which have in common the complementarity between the states of the high-side and low-side switches of each boost cell, i.e., when one of them is turned on, the other is turned off. A brief description of features distinguishing these techniques can be summarized below:

- The single surface sliding mode control (SS-SMC) produces two circuit structures and is obtained when the high-side switch of one boost cell is turned on and turned off simultaneously with the low-side switch of the other boost cell and the same for the other switches. In that case, it is possible to reduce the number of control signals. One control signal u_1 is sufficient to describe the circuit behavior. The corresponding model is obtained replacing u_2 by $1 - u_1$ in the previous model. The use of this coupled operation of the switches in the control law allows to use only the measurement of the output voltage and both inductor currents to ensure the desired behavior. Furthermore, only one hysteresis comparator enforces the sliding regime tracking the inner reference given by an outer controller which in turn enforces a pure sine-waveform behavior in the output voltage. Although optimal in terms of implementation and computational cost, this commutation method is limited to guarantee a single control objective: provide a high quality output voltage.
- The double surface sliding mode control (DS-SMC) produces four circuit structures and is obtained when the switches of one boost cell commutate completely independent of the switches of the other cell. This signifies that control of the cells is independent although the cells are interconnected through the load and share the connection to the input DC voltage. This is the common way to configure the control loops although it requires measurement of both capacitor voltages and both inductor currents. In addition, one hysteresis comparator is required per boost cell to track the reference given by the outer compensator operating on the capacitor voltage error. Although two

DC biased pure sine-waves are normally used as references, as discussed in the previous section, enriching their harmonic content allows considerably reducing the voltage stress of the switches. Therefore, this control approach is adopted to develop the contribution of this study.

Consider the following sliding surfaces

$$S(x) = \begin{bmatrix} S_1(x) \\ S_2(x) \end{bmatrix} = \begin{bmatrix} i_1(t) - i_{1e}(t) \\ i_2(t) - i_{2e}(t) \end{bmatrix} \tag{15}$$

where $i_{1e}(t)$ and $i_{2e}(t)$ are given current signals. By applying the invariance conditions $S(x) = 0$ and replacing in Equations (9) and (10), the equivalent controls are given by:

$$1 - u_{1_{eq}} = \frac{v_{in}(t) - L\frac{di_{1e}(t)}{dt}}{v_1(t)} > 0 \Rightarrow \frac{di_{1e}(t)}{dt} < \frac{v_{in}(t)}{L} \tag{16}$$

$$1 - u_{2_{eq}} = \frac{v_{in}(t) - L\frac{di_{2e}(t)}{dt}}{v_2(t)} > 0 \Rightarrow \frac{di_{2e}(t)}{dt} < \frac{v_{in}(t)}{L} \tag{17}$$

Replacing Equations (16) and (17) in Equations (12) and (13) and because on the surface $i_1(t) = i_{1e}(t)$ and $i_2(t) - i_{2e}(t)$, the following equations are obtained defining the ideal sliding dynamic of the inverter:

$$\frac{dv_1}{dt} = \frac{1}{C}\frac{v_{in}i_{1e}}{v_1} + \frac{1}{r_0 C}(v_2 - v_1) - \frac{L}{C}\frac{i_{1e}}{v_1}\frac{di_{1e}}{dt} \tag{18}$$

$$\frac{dv_2}{dt} = \frac{1}{C}\frac{v_{in}i_{2e}}{v_2} - \frac{1}{r_0 C}(v_2 - v_1) - \frac{L}{C}\frac{i_{2e}}{v_2}\frac{di_{2e}}{dt} \tag{19}$$

Now, consider that converter variables has incremental variations around one instantaneous operation point $v_1 = V_1 + \widetilde{v}_1$, $v_2 = V_2 + \widetilde{v}_2$, $i_{1e} = I_{1e} + \widetilde{i}_{1e}$, $i_{2e} = I_{2e} + \widetilde{i}_{2e}$, $\frac{di_{1e}}{dt} = I'_{1e} + \widetilde{i}'_{1e}$, $\frac{di_{2e}}{dt} = I'_{2e} + \widetilde{i}'_{2e}$, $v_{in} = V_{in} + \widetilde{v}_{in}$ and $r_0 = R_0 + \widetilde{r}_0$. By following the conventional linearization procedure preserving only first-order terms, it is obtained that:

$$\frac{d\widetilde{v}_1}{dt} = W_{11}\widetilde{i}_{1e} - LW_{12}\widetilde{i}'_{1e} - (CW_{11}W_{12} + b)\widetilde{v}_1 + b\widetilde{v}_2 + W_{12}\widetilde{v}_{in} - W_3\widetilde{r}_0 \tag{20}$$

$$\frac{d\widetilde{v}_2}{dt} = W_{21}\widetilde{i}_{2e} - LW_{22}\widetilde{i}'_{2e} - (CW_{21}W_{22} + b)\widetilde{v}_1 + b\widetilde{v}_1 + W_{22}\widetilde{v}_{in} + W_3\widetilde{r}_0 \tag{21}$$

$$\begin{aligned} W_{11} &= \tfrac{1}{CV_1}\left(V_{in} - LI'_{1e}\right) & W_{12} &= \tfrac{I_{1e}}{CV_1} & b &= \tfrac{1}{R_0 C} \\ W_{21} &= \tfrac{1}{CV_2}\left(V_{in} - LI'_{2e}\right) & W_{22} &= \tfrac{I_{2e}}{CV_2} & W_3 &= \tfrac{V_2 - V_1}{R_0^2 C} \end{aligned} \tag{22}$$

By applying the Laplace transform to Equations (20) and (21), Equations (23) and (24) are obtained with which the linear model of whole system can be represented using the block diagram in Figure 5. Transfer functions are defined using the polynomial Equations (25)–(30) considering $G_1(s) = B_1(s)/A(s)$, $G_2(s) = B_2(s)/A(s)$, $H_{12}(s) = C_1(s)/A(s)$, $H_{21}(s) = C_2(s)/A(s)$, $H_{v1}(s) = D_1(s)/A(s)$, $H_{v2}(s) = D_2(s)/A(s)$, $H_{r1}(s) = E_1(s)/A(s)$ and $H_{r2}(s) = E_2(s)/A(s)$.

$$V_1(s) = G_1(s)I_{1e}(s) + H_{12}(s)I_{2e}(s) + H_{v1}(s)V_{in}(s) - H_{r1}(s)R_0(s) \tag{23}$$

$$V_2(s) = G_2(s)I_{2e}(s) + H_{21}(s)I_{1e}(s) + H_{v2}(s)V_{in}(s) + H_{r2}(s)R_0(s) \tag{24}$$

$$(s) = s^2 + [C(W_{11}W_{12} + W_{21}W_{22}) + b]s + C^2 W_{11}W_{12}W_{21}W_{22} + bC(W_{11}W_{12} + W_{21}W_{22}) \tag{25}$$

$$B_1(s) = -LW_{12}s^2 + [W_{11} - LW_{12}(CW_{21}W_{22} + b)]s + W_{11}(CW_{21}W_{22} + b) \tag{26}$$

$$B_2(s) = -LW_{22}s^2 + [W_{21} - LW_{22}(CW_{11}W_{12} + b)]s + W_{21}(CW_{11}W_{12} + b) \tag{27}$$

$$C_1(s) = b(W_{21} - LW_{22}s) \qquad\qquad C_2(s) = b(W_{11} - LW_{12}s) \tag{28}$$

$$D_1(s) = W_{12}s + W_{12}(CW_{21}W_{12} + b) + bW_{22} \qquad D_2(s) = W_{22}s + W_{22}(CW_{11}W_{12} + b) + bW_{12} \tag{29}$$

$$E_1(s) = W_3(s + CW_{21}W_{22}) \qquad\qquad E_2(s) = W_3(s + CW_{11}W_{12}) \tag{30}$$

As it can be noted, the parameters of the plant transfer functions $G_1(s)$ and $G_2(s)$. have a high dependence not only on the operation point defined by the input voltage and the load, but also on the shape of the converter waveforms. In order to accurately model the behavior of these parameters, the inductor currents and their derivatives are analyzed in the next section through application of harmonic balance technique facilitating the synthesis of the simplest controller.

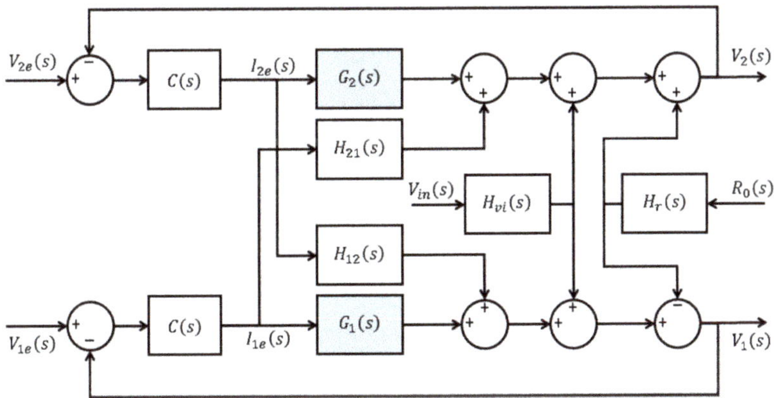

Figure 5. Block diagram of the outer voltage controllers.

3.3. Analysis of the Dynamic Behavior using Harmonic Balance

From the analysis in Section 2, the reference required to enforce the desired shape in the capacitor voltages has the following form:

$$v_{1e} = V_{dc} - V_{s1}\sin\omega t + V_{c2}\cos 2\omega t + V_{c4}\cos 4\omega t \tag{31}$$

$$v_{2e} = V_{dc} + V_{s1}\sin\omega t + V_{c2}\cos 2\omega t + V_{c4}\cos 4\omega t \tag{32}$$

The corresponding time derivatives are given by:

$$\frac{dv_{1e}}{dt} = -\omega V_{s1}\cos\omega t - 2\omega V_{c2}\sin 2\omega t - 4\omega V_{c4}\sin 4\omega t \tag{33}$$

$$\frac{dv_{2e}}{dt} = \omega V_{s1}\cos\omega t - 2\omega V_{c2}\sin 2\omega t - 4\omega V_{c4}\sin 4\omega t \tag{34}$$

Now, suppose that the stationary periodic behavior of the inductor currents can be approximated, in a satisfactory way by means of the following waveforms and its derivatives:

$$i_{1e} = I_{dc} - I_{s1}\sin\omega t - I_{c1}\cos\omega t + \sum_{n=2}^{4}(-1)^{n+1}[I_{sn}\sin n\omega t + I_{cn}\cos n\omega t] \tag{35}$$

$$i_{2e} = I_{dc} + I_{s1}\sin\omega t + I_{c1}\cos\omega t - \sum_{n=2}^{4}[I_{sn}\sin n\omega t + I_{cn}\cos n\omega t] \tag{36}$$

$$\frac{di_{e1}}{dt} = \omega I_{c1} \sin \omega t - \omega I_{s1} \cos \omega t + \omega \sum_{n=2}^{4} n\left[(-1)^n I_{cn} \sin n\omega t + (-1)^{n+1} I_{sn} \cos n\omega t\right] \tag{37}$$

$$\frac{di_{2e}}{dt} = -\omega I_{c1} \sin \omega t + \omega I_{s1} \cos \omega t + \omega \sum_{n=2}^{4} n\left[I_{cn} \sin n\omega t - I_{sn} \cos n\omega t\right] \tag{38}$$

On the other hand, expressions Equations (18) and (19) can be interpreted as power balance constraints for the boost inverter cells. Then, they can be rewritten as follows:

$$Li_{1e}\frac{di_{1e}}{dt} + Cv_{1e}\frac{dv_1}{dt} - \left(v_{in}i_{1e} + \frac{v_1 v_{oe}}{r_0}\right) = 0 \tag{39}$$

$$Li_{2e}\frac{di_{2e}}{dt} + Cv_2\frac{dv_2}{dt} - \left(v_{in}i_{2e} - \frac{v_2 v_{0e}}{r_0}\right) = 0 \tag{40}$$

Summing Equations (39) and (40), the power balance equation for the complete inverter can be written as:

$$L\left(i_{e2}\frac{di_{e2}}{dt} + i_{e1}\frac{di_{e1}}{dt}\right) + C\left(v_{e1}\frac{dv_{e1}}{dt} + v_{e2}\frac{dv_{e2}}{dt}\right) - \left[v_{in}(i_{e1} + i_{e2}) - \frac{v_{eo}^2}{r_0}\right] = 0 \tag{41}$$

By replacing Equations (31)–(38) into Equation (41) and applying the harmonic balance technique, it is obtained a set of nonlinear algebraic equations, whose unknowns are the amplitudes of the harmonic components of the currents. It can be compactly written as:

$$F_B\left(I_{dc}, , I_{s1}, I_{c1}, I_{s2}, I_{c2}, I_{s3}, I_{c3}, I_{s3}, I_{c4}, V_{dc}, V_{s1}, V_{c2}, V_{c4}\right) = 0 \tag{42}$$

where the components of F_B are expressed by:

$$
\begin{aligned}
F_{dc} &= 2V_{in}I_{dc} - \frac{2V_{s1}^2}{r_0} \\
F_{s2} &= \omega L\left(I_{s1}^2 - I_{c1}^2 + 4I_{dc}I_{c2} + 2I_{s1}I_{s3} + 2I_{c1}I_{c3} - 2I_{s2}I_{s4} - 2I_{c2}I_{c4}\right) + \omega C\left(V_{s1}^2 - 4V_{dc}V_{c2} - 2V_{c2}V_{c4}\right) + 2V_{in}I_{s2} \\
F_{c2} &= 2\omega L\left(I_{s1}I_{c1} - 2I_{dc}I_{s2} + 2I_{s1}I_{c3} - 2I_{s3}I_{c1} + 2I_{s4}I_{c2} - 2I_{s2}I_{c4}\right) - \left(-2V_{in}I_{c2} + \frac{2V_{s1}^2}{r_0}\right) \\
F_{s4} &= 2\omega L\left(I_{s2}^2 - I_{c2}^2 - 2I_{s1}I_{s3} + 2I_{c1}I_{c3} + 4I_{dc}I_{c4}\right) + 2\omega C\left(-V_{c2}^2 - 4V_{dc}V_{c4}\right) + 2V_{in}I_{s4} \\
F_{c4} &= 4\omega L\left(I_{s2}I_{c2} - I_{s3}I_{c1} - I_{s1}I_{c3} - 2I_{dc}I_{s4}\right) + 2V_{in}I_{c4} \\
F_{s6} &= 3\omega L\left(I_{s3}^2 - I_{c3}^2 + 2I_{s2}I_{s4} - 2I_{c2}I_{c4}\right) - 6\omega CV_{c2}V_{c4} \\
F_{c6} &= 6\omega L\left(I_{s3}I_{c3} + I_{s2}I_{c4} + I_{s4}I_{c2}\right) \\
F_{s7} &= 7\omega L\left(-I_{s3}I_{s4} + I_{c3}I_{c4}\right) \\
F_{c7} &= 7\omega L\left(-I_{s3}I_{c4} - I_{s4}I_{c3}\right) \\
F_{s8} &= 4\omega L\left(I_{s4}^2 - I_{c4}^2\right) - 4\omega CV_{c4}^2 \\
F_{c8} &= 8\omega L I_{s4}I_{c4}
\end{aligned}
\tag{43}
$$

Numeric evaluation of this nonlinear equation system for the entire range of v_{in} and r_0 allows to obtain the shape of the inductor currents and their derivatives which beside to the desired voltage references allow to analyze the dynamic behavior of the system. Solutions are obtained by using the function *lsqnonlin* of MATLAB considering the input–output ranges listed in Table 1 and parameters in Table 3. For the subsequent analysis, recall that harmonic components of the capacitor voltages for a given amplitude V_m of the desired output voltage $v_0(t)$ are expressed by $V_{s1}0.5V_m$, $V_{c2}0.18V_m$ and $V_{c4}0.018V_m$. The dC component of the voltage references is computed some volts higher than the minimum permissible value: $V_{dc} = V_{in} + 0.338V_m + V_+$.

Table 1. Range of operation of the converter used for numeric analysis.

Parameter	Symbol	Minimum	Maximum	Unities
Input voltage	v_{in}	125%–25%	125 + 25%	V
Output voltage amplitude (Std. 1)	v_m	120 $\sqrt{2}$		V
Output frequency (Std. 1)	f	60		Hz
Output power (Std. 1)	P_o	24	240	W
Resistive load (Std. 1)	r_o	600	60	Ω
Output voltage amplitude (Std. 2)	v_m	220 $\sqrt{2}$		V
Output frequency (Std. 2)	f	50		Hz
Output power (Std. 2)	P_o	22	220	W
Resistive load (Std. 2)	r_o	2200	220	Ω

By evaluating the periodic terms of Equations (16) and (17), it is possible to observe that their values are always lower than any value of the input voltage in the range of operation of the converter. Figure 6 shows three cycles of these terms evaluated for 30 coordinates into the range of input voltage and output load including extreme values. A coincidence of the produced waveforms around zero is recognized. From these results, it is easy to conclude, that operating with constant voltage and load values, this condition will never be violated. However, during transient response to input voltage or power load disturbances, the value of the derivative can considerably increase enforcing the loss of the sliding regime. As we will see later, maintaining the system in the sliding regime can be ensured saturating the derivative of the currents.

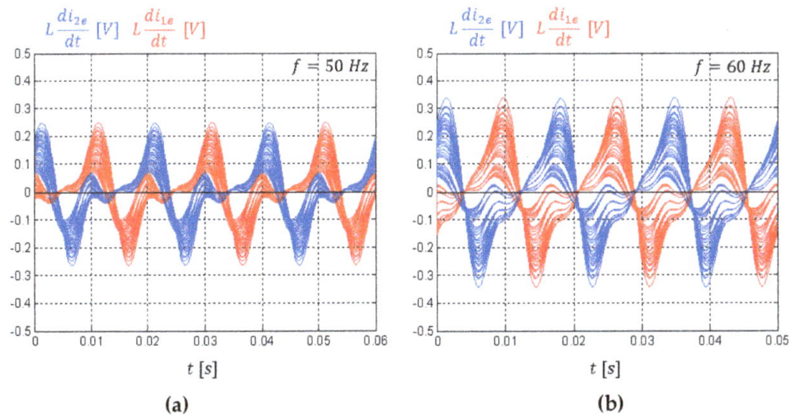

Figure 6. Evaluation of three cycles of the waveforms of the inductor voltages for the entire range of input voltage at full load. (**a**) Standard 220 V @ 50 Hz and (**b**) standard 120 V @ 60 Hz.

Figure 7 depicts the inductor current waveforms evaluated for the same 30 coordinates into the set of operation conditions of the converter. As it can be observed, when the inductor current of one cell takes negative values, the current in the other one takes positive values. In some few cases and for short intervals close to the end of each half-period, both currents can take negative values simultaneously. A coincidence around zero is also observed for this variable.

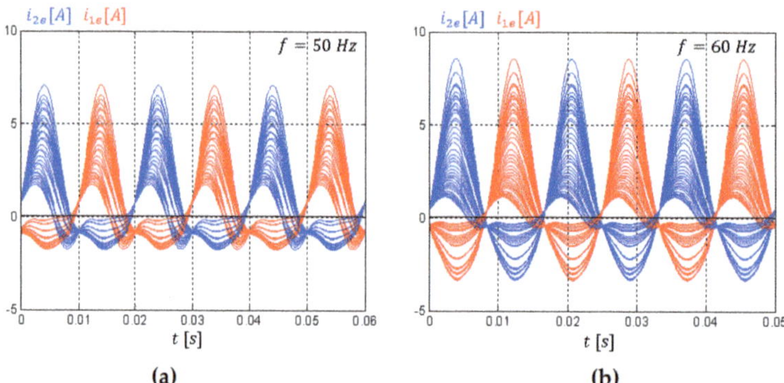

Figure 7. Evaluation of three cycles of the waveforms of the inductor currents for the entire ranges of input voltage and output load. (a) Standard 220 V @ 50 Hz and (b) standard 120 V @ 60 Hz.

Figure 8 shows the capacitor voltage waveforms evaluated for the whole range of input voltage of the converter. Please note that impose these voltage waveforms is one of the control objectives and then the shape of the voltages must be independent of the power load. It is possible to observe how independent of the output frequency, the waveforms show coincidence at the end of each half-cycle taking values in the vicinity of $2V_{in}$.

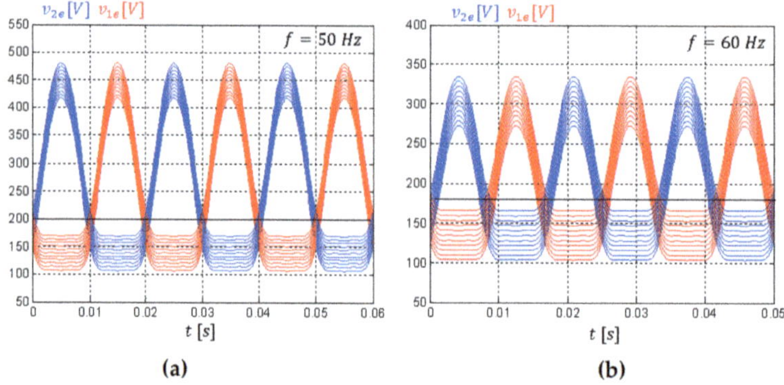

Figure 8. Evaluation of three cycles of the waveforms of the capacitor voltages for the entire range of input voltage at full load. (a) Standard 220 V @ 50 Hz and (b) standard 120 V @ 60 Hz.

From the results of the previous analysis, it seems to be reasonable to evaluate the parameters of the polynomial Equations (25) and (27) to obtain $G_2(s)$ around the average point Equation (44) which results in the reference current to capacitor voltage transfer functions given by Equation (45).

$$\left(i_{1e}, i_{2e}, v_1, v_2, L\frac{di_{1e}}{dt}, L\frac{di_{2e}}{dt}, v_{in}, r_0\right) = (0, 0, 2V_{in}, 2V_{in}, 0, 0, V_{in}, R_0) \tag{44}$$

$$G_1(s) = \frac{s+b}{2C(s+2b)s} \qquad\qquad G_2(s) = \frac{s+b}{2C(s+2b)s} \tag{45}$$

The resulting model shows that system dynamics is not asymptotically stable nor unstable. Around the linearization point, an integral effect can be observed, but this effect diminishes as we move away from it. The following as conclusions can be considered for the controller design stage.

1. Without an outer loop, having that the average value of the currents is positive, the voltages of the capacitors will increase until the permitted physical limits;

2. Using a proportional controller, the stability is ensured, but the references cannot be accurately tracked. The error is greater for instantaneous operation points furthest from linearization point;
3. A proportional-integral (PI) controller with adequate parameters can provide an accurate tracking of the periodic references for all operation points;
4. The resulting transfer functions in Equation (45) show no effect of the input and output voltages but include the term b which represents the influence of the power load.

3.4. Outer Voltage Controllers

The tracking of the voltage references is guaranteed by the outer loops with PI voltage compensators defined by:

$$\begin{cases} s_x(t) = sat_{u_0}\left\{K_p\left[\dfrac{d(v_x(t)-v_{xe}(t))}{dt} + \alpha(v_x(t) - v_{xe}(t))\right]\right\} \\ i_{xe}(t) = \int_{-\infty}^{t} s_x(\tau)d\tau \quad x = 1,2 \end{cases} \tag{46}$$

where $sat_{u_0}(x)$ is the classical symmetrical saturation function [17] having $u_0 = \frac{v_{in}(t)}{L}$ defining its limits. K_p and α are positive design parameters, K_p being the proportional gain and αK_p the integral gain. When the signal $s_x(t)$ does not saturate, the controller is a classical PI controller whose expression is:

$$i_{xe}(t) = K_p(v_x(t) - v_{xe}(t)) + \alpha K_p \int_{-\infty}^{t} (v_x(\tau) - v_{xe}(\tau))d\tau, \quad x = 1,2$$

The presence of saturation function ensures that the overall controlled system is asymptotically stable. This can be deduced from the stability result developed in the previous section, simply by noting that functions $a_x(t), x = 1,2$, remain positive if condition Equation (15) is satisfied, that is, if:

$$\frac{di_{xe}(t)}{dt} < \frac{v_{in}(t)}{L}, \quad x = 1,2$$

Equation (15) being also a necessary condition for the existence of a sliding regime. In such situation, for all positive K_p and α, we will have:

$$\lim_{t\to\infty} v_1(t) = v_{1e}(t) \text{ and } \lim_{t\to\infty} v_2(t) = v_{2e}(t)$$

For ease implementation of the proposed controller, we can remark that an expression of the outer control can be formulated with a saturation function whose limit is constant and equal to a positive real number s_0 (i.e., independent of time) chosen to facilitate the controller implementation. An alternate expression with such a property could be:

$$\begin{cases} s_x(t) = \dfrac{v_{in}(t)}{Ls_0} sat_{s_0}\left\{\dfrac{Ls_0}{v_{in}(t)}K_p\left[\dfrac{d(v_x(t)-v_{xe}(t))}{dt} + \alpha(v_x(t) - v_{xe}(t))\right]\right\} \\ i_{xe}(t) = \int_{-\infty}^{t} s_x(\tau)d\tau \quad x = 1,2 \end{cases} \tag{47}$$

Figure 9 represents the block-diagram of the proposed saturated PI controller.

To simplify the controller structure and then its implementation, if there exists a value $v_{in_{min}} > 0$ such that for all t, $v_{in}(t) > v_{in_{min}}$, it is possible to drop the measurement of $v_{in}(t)$ and replace it by $v_{in_{min}}$ in the control expression. It is also possible to use a non-saturated PI controller if the converter was adequately designed to track the reference signals of interest (i.e., in a way guaranteeing that the control never saturates).

To end with the outer loop, the choice of constants K_p and α have an influence on the transient behavior of the controlled system, but asymptotic stability is always ensured. Even if asymptotic stability is a necessary condition, it is not sufficient in practice. A good performance level is often needed. The parameters K_p and αK_p can be selected as done classically for a nonsaturated PI controller. However, it is important to remark that while the stability will be preserved, the performance associated with such a choice will only be guaranteed in the zone of linearity of the saturated PI controller. When the controller will saturate, except stability, no performance level can be guaranteed.

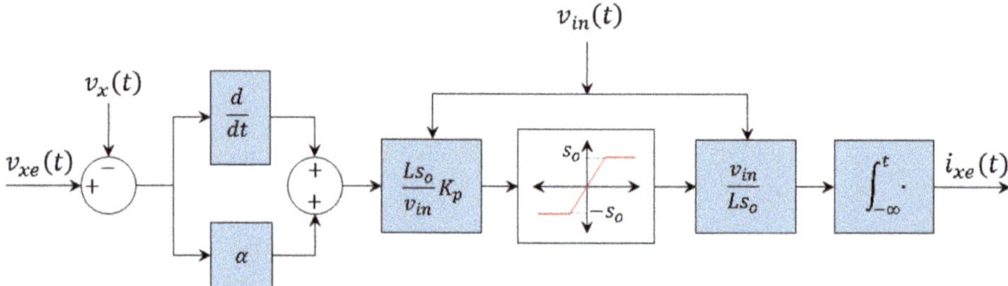

Figure 9. Block diagram of the outer voltage controllers.

4. Validation Results via Simulation

To validate the accurateness of the mathematical procedures and the correct operation of the proposed control, two simulation tests were built in PSIM software, one for standard of 120 V @ 60 Hz (Simulation test 1) and another for 220 V @ 50 Hz (Simulation test 2). Comparison of voltage stress in semiconductors with respect to the SS-SMC presented in [4] is also performed. The simulation step time was configured to be 2.60417×10^{-7} s for the test 1 and 3.125×10^{-7} s for the test 2. The parameters of the power converter and its control are listed in Tables 2 and 3. Parasitic resistances of elements were obtained from datasheets: Equivalent series resistances in inductors and capacitors (R_L and R_C, respectively) and Drain-Source on–resistance in MOSFETs (R_M). Considering that the proposed control can be entirely implemented with analog electronics, digitalization effects were only considered for reference generation. The generation method presented in [18] was considered, where signals are produced by using a look-up table storing 128 samples of 10 bits for a cycle of the output signal. It is worth mentioning that these features are typical of low cost microcontroller.

Table 2. Control parameters used in simulations.

Parameter	Symbol	Value (60 Hz)	Value (50 Hz)	Unities
Proportional gain	K_p	0.3	0.2	
Integral gain	αK_p	25×10^{-6}	20×10^{-6}	
Saturation limit	s_0	500×10^3	500×10^3	A/s
Hysteresis band width	2δ	2	2	A
Cutoff frequency LPF	f_{c1}	240	200	Hz

Table 3. Parameters of the power converter in simulations.

Element	Manuf./Reference	Parameter	Symbol	Value	Unities
Capacitors	KEMET [19]	Capacitance	C	9	μF
		Series resistance	R_C	8.3	mΩ
Inductors	Bourns [20]	Inductance	L	120	μH
		Series resistance	R_L	28	mΩ
MOSFETs	ROHM [21]	On-resistance	R_M	196	mΩ

In both simulation tests, five points are enforced to assess the stationary and transient behavior. Selected conditions and time intervals are listed in Table 4. Sudden transitions are enforced during the load changes while ramp type transitions with intervals of 3 ms are applied for input voltage changes.

Table 4. Operation conditions and time intervals used in simulation tests.

Convention	Input Voltage	Power Load	Test 1 (60 Hz)	Test 2 (50 Hz)
Operation condition 1	125.0 V	100%	0.10–0.15 s	0.10–0.16 s
Operation condition 2	93.75 V	100%	0.15–0.20 s	0.16–0.22 s
Operation condition 3	156.2 V	100%	0.20–0.25 s	0.22–0.28 s
Operation condition 4	156.2 V	25%	0.25–0.30 s	0.28–0.34 s
Operation condition 5	93.75 V	25%	0.30–0.35 s	0.34–0.40 s

4.1. Simulation Test 1 (American Standard 120 V @ 60 Hz)

Figure 10 presents the waveforms at the output of the converter (voltage and current), the capacitor voltages and the inductor currents for the simulation test 1. As it can be observed, the output signal accurately track the high quality sinusoidal in both stationary and transient regimes. After disturbances, the AC component of the capacitor voltages remains unchanged while its average value adapts to the input voltage. Table 5 summarizes the obtained THD (lower than 1%) and RMS error (lower than 0.2%) demonstrating the high performance of proposed control in the five selected operation points.

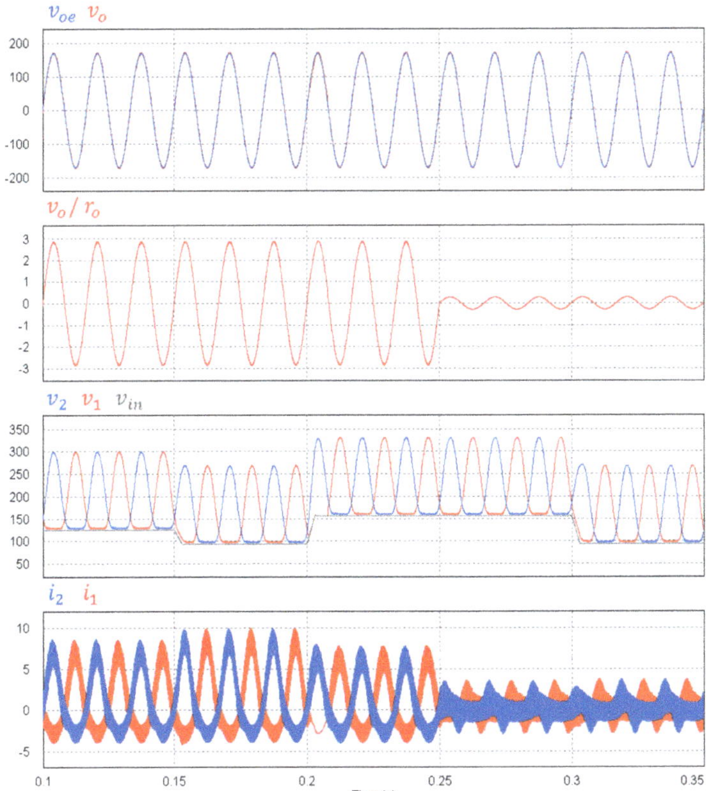

Figure 10. Simulated results during Experiment 1 using the proposed DS-SMC.

Table 5. Output voltage quality for simulation test 1 (120 V @ 60 Hz).

Convention	Time Interval	THD (%)	RMS Error (%)
Operation condition 1	0.10–0.15 s	0.70	0.06
Operation condition 2	0.15–0.20 s	0.70	0.06
Operation condition 3	0.20–0.25 s	0.71	0.07
Operation condition 4	0.25–0.30 s	0.71	0.08
Operation condition 5	0.30–0.35 s	0.66	0.10

Figure 11 shows a comparison between the DS–SMC approach developed in this work and the SS-SMC approach developed in [4]. As it can be noted, the voltage of the capacitors which is the same voltage applied to open semiconductors in each boost cell is always lower for the DS-SMC. It is relevant to mention that two features of the proposed control are responsible of the improvement: (a) the harmonic content included in the capacitor voltages and (b) their average component. The DC component cannot be accessed using the SS-SMC approach while using the proposed DS-SMC this component allows to enforce the minimum value of the output capacitors to be almost equal to the input voltage. In this simulation test $V_{dc} = V_{in} + 0.338V_m + V_+$ being V_+ settled to 5 V.

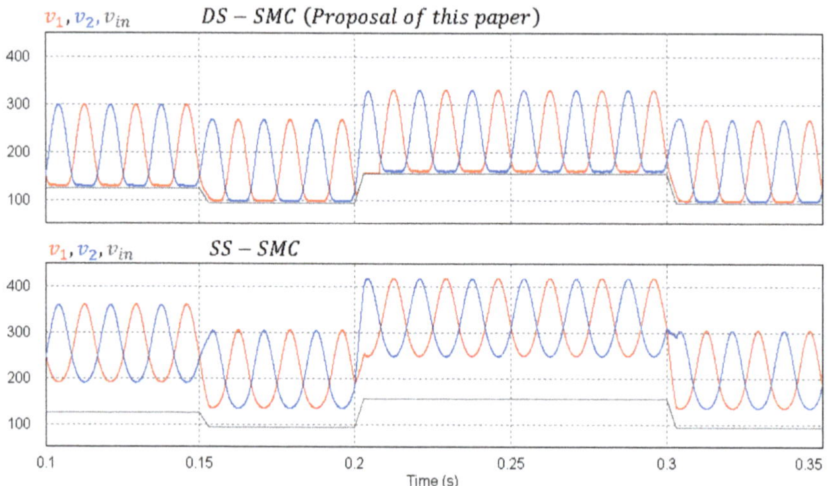

Figure 11. Simulated semiconductor voltages comparing SS-SMC and DS-SMC for simulation test 1.

4.2. Simulation Test 2 (European Standard 220 V @ 50 Hz)

Figure 12 presents the waveforms at the converter for the simulation test 2. It is possible to confirm that the output signal accurately track the high quality sinusoidal. Table 6 summarizes the obtained THD and RMS error which show values lower than 1% and 0.2%, respectively. It is worth mentioning that the required voltage gain increases around 50% without affecting the performance of the control which use the same parameters.

Table 6. Output voltage quality for simulation test 2 (220 V @ 50 Hz).

Convention	Time Interval	THD (%)	RMS Error (%)
Operation condition 1	0.10–0.16 s	0.76	0.08
Operation condition 2	0.16–0.22 s	0.76	0.10
Operation condition 3	0.22–0.28 s	0.76	0.08
Operation condition 4	0.28–0.34 s	0.77	0.09
Operation condition 5	0.34–0.40 s	0.74	0.11

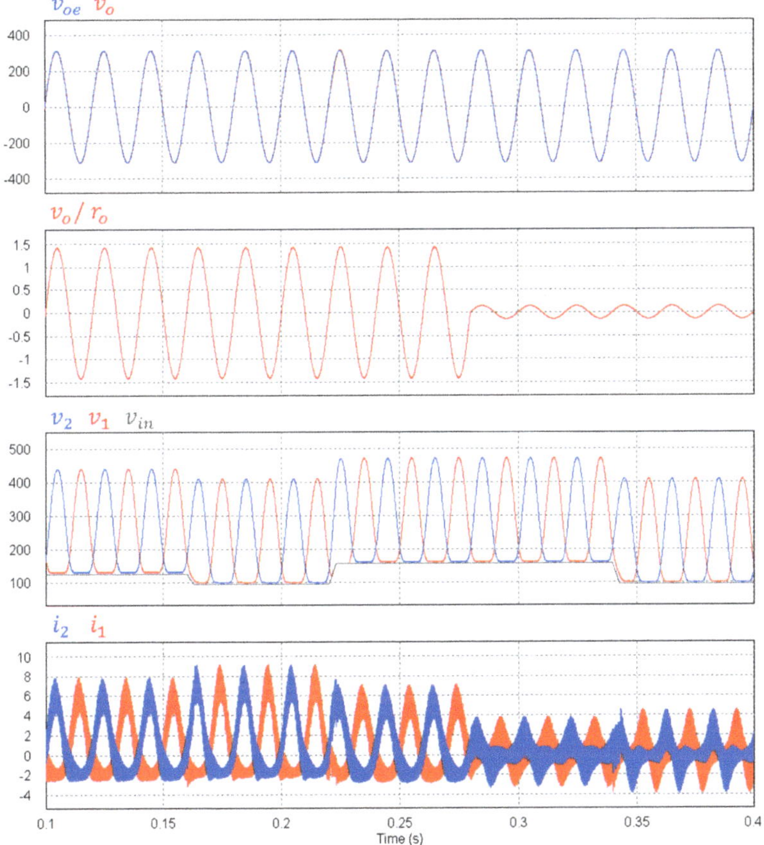

Figure 12. Simulated waveforms during experiment 2 using the proposed DS-SMC.

Equal to the previous test, the DS–SMC and SS-SMC approaches were compared. Again, the voltage applied to semiconductors is always lower for the DS-SMC. The expression $V_{dc} = V_{in} + 0.338V_m + V_+$ being V_+ settled to 5 V is used to define input voltage. In addition, similar to the previous case, the alleviation of the semiconductor voltage stress is at least of 30 V and becomes up to 100 V. A more accurate analysis is presented in the next subsection.

4.3. Semiconductor Voltage Stress Comparison

A further analysis was done by reviewing the maximum, minimum and average semiconductor voltage stress for inverter operating with the two studied standards using the same simulations producing Figures 11 and 13. In addition to the two methods compared in these figures, the proposed control without adding the harmonic components in the voltage references is also assessed. In Table 7, the minimum voltage corresponds with intervals in which the input voltage is minimum (125 V–25%), the maximum voltage corresponds to the intervals in which the input voltage is maximum (125 V + 25%) and the average is computed considering the complete simulation interval for both standards.

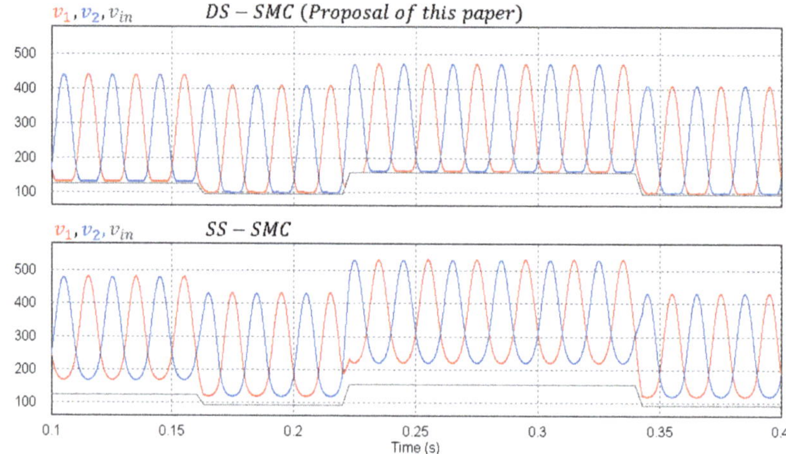

Figure 13. Simulated semiconductor voltages comparing SS-SMC and DS-SMC for simulation test 2.

Table 7. Comparison of control methods regarding semiconductor voltage stress.

Control Method	Standard 1 (120 V @ 60 Hz)			Standard 1 (220 V @ 50 Hz)		
	Min (V)	Max (V)	Avg (V)	Min (V)	Max (V)	Avg (V)
SS-SMC [4]	136	418	254	120	533	290
DS-SMC (Sine references)	97	331	209	97	472	285
DS-SMC (Modified references)	97	331	182	97	472	235

Results demonstrate how the voltage stress of the semiconductors reduces considerably by employing the DS-SMC proposed in this study (Separate voltage control loops minimizing the DC-bias of the references) even without additional harmonic components into the voltage references. Beyond that, results also demonstrate conclusively how the addition of the two harmonic components in the references further improve this feature.

5. Conclusions

In this study, the sliding mode control technique was effectively applied to accomplish two main objectives in the operation of the boost inverter: (a) provide a high quality output voltage and (b) minimize the required voltage gain in the boost cells to alleviate semiconductor voltage stress. Different from all previously published works related to the control of this converter, the voltage of the capacitors are enforced to have additional optimal values of double and fourth harmonic terms. The objectives are accomplished by means of two independent multiloop controllers involving an inner loop of sliding mode control implemented using hysteresis comparators and PI compensators ensuring the tracking of harmonic enriched references. Additionally, having a degree of freedom to modify the average value of the capacitor voltages, an adaptive feed-forward loop was integrated helping to further reduce the required gain in the converter cells. A complete study of the stationary behavior of the converter was developed to obtain a simple model of its dynamics by using the harmonic balance technique. A PI controller with saturation of the output derivative allowed to ensure the sliding regime of the inner loop which is also an innovative feature of the proposed control scheme. The obtained THD is lower than 0.8% and the regulation of the RMS value is lower than 0.2% in the entire range of operation of the converter for both standards used as case study. The alleviation of the semiconductor voltage stress is very important also for both analyzed standards.

From the results of the present work, a new control scheme is being developed using a multiple input multiple output perspective of the problem, this allowing a better action on the coupling dynamics of the boost cells. Prospective work involves experimental validation using a laboratory prototype.

Author Contributions: All authors contributed equally to all the stages of the research process and the preparation of the manuscript. All authors have read and agreed to the published version of the manuscript.

Funding: This research was developed with the partial support of the Ministerio de Ciencia, Tecnología e Innovación de Colombia (MinCiencias) under contract 018–2016 and the HisPaliS project supported by the Agence Nationale de la Recherce de France (ANR).

Conflicts of Interest: The authors declare no conflicts of interest.

References

1. Caceres, R.O.; Barbi, I. A boost DC-AC converter: Analysis, design, and experimentation. *IEEE Trans. Power Electron.* **1999**, *14*, 134–141. [CrossRef]
2. Sanchis, P.; Ursaea, A.; Gubia, E.; Marroyo, L. Boost DC-AC inverter: A new control strategy. *IEEE Trans. Power Electron.* **2005**, *20*, 343–353. [CrossRef]
3. Cortes, D.; Vazquez, N.; Alvarez-Gallegos, J. Dynamical Sliding-Mode Control of the Boost Inverter. *IEEE Trans. Ind. Electron.* **2009**, *56*, 3467–3476. [CrossRef]
4. Flores-Bahamonde, F.; Valderrama-Blavi, H.; Bosque-Moncusi, J.M.; García, G.; Martínez-Salamero, L. Using the sliding-mode control approach for analysis and design of the boost inverter. *IET Power Electron.* **2016**, *9*, 1625–1634. [CrossRef]
5. Wai, R.; Lin, Y.; Liu, Y. Design of Adaptive Fuzzy-Neural-Network Control for a Single-Stage Boost Inverter. *IEEE Trans. Power Electron.* **2015**, *30*, 7282–7298. [CrossRef]
6. Abeywardana, D.B.W.; Hredzak, B.; Agelidis, V.G. A Fixed-Frequency Sliding Mode Controller for a Boost-Inverter-Based Battery-Supercapacitor Hybrid Energy Storage System. *IEEE Trans. Power Electron.* **2017**, *32*, 668–680. [CrossRef]
7. Jha, K.; Mishra, S.; Joshi, A. High-Quality Sine Wave Generation Using a Differential Boost Inverter at Higher Operating Frequency. *IEEE Trans. Ind. Appl.* **2015**, *51*, 373–384. [CrossRef]
8. Zhu, G.; Xiao, C.; Wang, H.; Tan, S. Closed-loop waveform control of boost inverter. *IET Power Electron.* **2016**, *9*, 1808–1818. [CrossRef]
9. Abeywardana, D.B.W.; Hredzak, B.; Agelidis, V.G. An Input Current Feedback Method to Mitigate the DC-Side Low-Frequency Ripple Current in a Single-Phase Boost Inverter. *IEEE Trans. Power Electron.* **2016**, *31*, 4594–4603. [CrossRef]
10. Abeywardana, D.B.W.; Hredzak, B.; Agelidis, V.G. A Rule-Based Controller to Mitigate DC-Side Second-Order Harmonic Current in a Single-Phase Boost Inverter. *IEEE Trans. Power Electron.* **2016**, *31*, 1665–1679. [CrossRef]
11. Zhao, W.; Lu, D.D.; Agelidis, V.G. Current Control of Grid-Connected Boost Inverter with Zero Steady-State Error. *IEEE Trans. Power Electron.* **2011**, *26*, 2825–2834. [CrossRef]
12. Jang, M.; Agelidis, V.G. A Minimum Power-Processing-Stage Fuel-Cell Energy System Based on a Boost-Inverter with a Bidirectional Backup Battery Storage. *IEEE Trans. Power Electron.* **2011**, *26*, 1568–1577. [CrossRef]
13. Abeywardana, D.B.W.; Acuna, P.; Hredzak, B., Aguilera, R.P.; Agelidis, V.G. Single-Phase Boost Inverter-Based Electric Vehicle Charger with Integrated Vehicle to Grid Reactive Power Compensation. *IEEE Trans. Power Electron.* **2018**, *33*, 3462–3471. [CrossRef]
14. Jang, M.; Ciobotaru, M.; Agelidis, V.G. A Single-Phase Grid-Connected Fuel Cell System Based on a Boost-Inverter. *IEEE Trans. Power Electron.* **2013**, *28*, 279–288. [CrossRef]
15. Jang, M.; Ciobotaru, M.; Agelidis, V.G. Design and Implementation of Digital Control in a Fuel Cell System. *IEEE Trans. Ind. Inf.* **2013**, *9*, 1158–1166. [CrossRef]
16. Tang, Y.; Bai, Y.; Kan, J.; Xu, F. Improved Dual Boost Inverter with Half Cycle Modulation. *IEEE Trans. Power Electron.* **2017**, *32*, 7543–7552. [CrossRef]
17. Mellodge, P. Chapter 4-Characteristics of Nonlinear Systems. In *A Practical Approach to Dynamical Systems for Engineers*; Woodhead Publishing: Cambridge, UK, 2016; pp. 215–250, ISBN 978-0-08-100202-5.

18. Lopez-Santos, O.; Tilaguy-Lezama, S.; Garcia, G. Adaptive Sampling Frequency Synchronized Reference Generator for Grid Connected Power Converters. *Commun. Comput. Inf. Sci.* **2018**, *915*, 573–587. [CrossRef]
19. KEMET. Datasheet Part Number C4A MKP SERIES, Box Capacitors PCB Applications, 1. 2017. Available online: https://content.kemet.com/datasheets/F3303_C4A.pdf (accessed on 1 July 2019).
20. Bourns. Datasheet High Current Chokes, 1–2. 2014. Available online: https://www.bourns.com/docs/Product-Datasheets/1140_series.pdf (accessed on 1 July 2019).
21. ROHM. R6020KNZ1, Nch 600V 20 A Power MOSFET Datasheet, 1–14 2019. Available online: https://fscdn.rohm.com/en/products/databook/datasheet/discrete/transistor/mosfet/r6020knz4c13-e.pdf (accessed on 1 July 2019).

Article

Improvement of Power Converters Performance by an Efficient Use of Dead Time Compensation Technique

Sheeraz Iqbal [1,2,*], Ai Xin [1], Mishkat Ullah Jan [1], Mohamed Abdelkarim Abdelbaky [3,4], Haseeb Ur Rehman [1], Salman Salman [1], Muhammad Aurangzeb [1], Syed Asad Abbas Rizvi [1] and Noor Ahmad Shah [1]

[1] School of Electrical and Electronic Engineering, North China Electric Power University, Beijing 102206, China; aixin@ncepu.edu.cn (A.X.); engrmishkat@gmail.com (M.U.J.); haseebups@gmail.com (H.U.R.); salman.ali1050@gmail.com (S.S.); maurangzaib42@gmail.com.com (M.A.); saarizvi.12@outlook.com (S.A.A.R.); noorahmadshah2k14@gmail.com (N.A.S.)

[2] Department of Electrical Engineering, University of Azad Jammu and Kashmir, Muzaffarabad 13100, Pakistan

[3] School of Control and Computer Engineering, North China Electric Power University, Beijing 102206, China; m_abdelbaky@ncepu.edu.cn

[4] Department of Electrical Power and Machines, Faculty of Engineering, Cairo University, Giza 12613, Egypt

* Correspondence: engrsheeraziqbal@gmail.com

Received: 3 April 2020; Accepted: 28 April 2020; Published: 29 April 2020

Abstract: The advent of renewable energy resources and distributed energy systems herald a new set of challenges of power quality, efficient distribution, and stability in the power system. Furthermore, the power electronic converters integration has been increased in interfacing alternate energy systems and industries with the transmission and distribution grids. Owing to the intermittency of renewable energy resources and the application of power electronic converters the power distribution faces peculiar challenges. The dead-time effects are among the main challenges, which leads to the distortion of third harmonics, phase angle, torque pulsation, and induction motor current, causing severe quality problems for power delivery. To tackle these problems, this paper proposes a novel dead time compensation technique for improving the power quality parameters and improving the efficiency of power converters. The proposed model is simulated in MATLAB and the parametric equations are plotted against the corresponding parametric values. Furthermore, by implementing the proposed strategy, significant improvements are attained in the torque pulsation, speed, and total harmonic distortion of the induction motor. The comparisons are drawn between with and without dead time compensation technique, the former shows significant improvements in all aspects of the power quality parameters and power converters efficiency.

Keywords: industrial microgrid; dead-time compensation; power quality; variable frequency drive; third harmonic distortion; induction motor

---

1. Introduction

The problems at the distribution side of electrical power systems are categorized under the umbrella of power quality problems consisting of distortion in phase angle, voltage waveforms, fundamental current, as well as frequency. The definition of power quality has been used to include several issues related to a power supply such as voltage and current quality, the stability of power supply, quality of the overall system, and efficiency of supply and consumption of power [1]. These problems primarily stem from the non-ideal characteristics of power electronic equipment. The power providers are warranted to ensure stable, smooth, and safe power supply conforming to a pure sinusoidal voltage signal to the end-user [2]. At present-day, with the worldwide energy crisis becoming increasingly conspicuous and environmental pollution becoming increasingly serious, worldwide renewable generation technologies have been developed rapidly [3–5]. However, the introduction of power electronics, necessitated by the presence of renewable energy sources (RES) at the distribution junction inflicts distortions in current and voltage causing difficulties in ensuring smooth electric supply [6]. Contemporary research is focused on finding novel ways to achieve the standards of power quality and with the forecast of incessant RES growth will be a major challenge for years to come. The adverse impacts of less than ideal power quality are amply documented in [7]. Almost all of these benchmarks are application-based, there have been almost tens of rules established by the Institute of European Commission (IEC) for the power quality. Although the most important power quality standard is IEC 61000-2-2, which ensures that the voltage harmonic levels in the power system do not surpass the compatibility levels [8].

Non-linear power electronic products have been finding a greater market in the residential and industrial sectors owing to their cost-effectiveness and high efficiency. Variable frequency drives (VFDs), power factor (PF) correction equipment, and switch-mode supplies (SMPSs) improve the overall system efficiency [9]. Additionally, industrial application for power converters is also increasing. This has shifted the overall research focus on the development of state-of-the-art power converters [10]. Pulse width modulation (PWM) and voltage source inverter (VSI) are generally installed for motor drives. Ideally, the turn-off-on of the two power devices at each leg of an inverter is complimentary. However, in the practical application, the time delay in the turn-off of one and turn-on of the other device may lead to short-circuiting of DC-link due to momentary simultaneous conduction [11]. To address this problem a blank duration, called dead-time, is inserted between the switching on of the one device and off of the other ensuring safe operation [12]. The effective voltage is affected by the dead time at lower frequencies, which further distorted the inverter output voltage and results in additional components of low-order harmonics. Moreover, this also causes distortion in the current waveforms [13]. These effects necessitate devising novel dead-time compensation strategies detailed in [14]. The strategies are categorized in two broad types: (1) feed-forward compensation entailing calculation of error stemming from the dead time and the concomitant forward voltage drop, and their subsequent compensation through control algorithms; and (2) stringent observation of disturbance magnitude and the subsequent proportionate compensation.

The first category of compensation strategies is contingent on the detection of current polarity-made difficult by high-frequency disturbances, and a phenomenon called "zero-current-clamp" [15]. Current polarity detection circuits are undesirable due to complicated structures and extra cost [16]. The low pass interference filtering strategy is rendered inefficient due to the phase lag they generate inducing new errors. Consequently, the current research is focused more on uncovering alternative means of current polarity detection, and interference filtering devoid of the phase lag problem. One such strategy involves the use of current reference value to offset the influence of clamp and avoid the repeated near-zero crossing of the actual sampled current [17]. This strategy, however, suffers from the drawback of potential error between the actual and reference current as well as its suitability for only closed-loop control of current. Some have devised a strategy of calculating the current polarity angle and place a high significance on the angle between the current vector and rotor flux angle [18].

Similarly, some have employed the method of dead-time and forward voltage drop induced error calculation in off-line or on-line mode, followed by the addition of the error to power devices' driving pulses [19]. In [20], the dead-time effect is aptly compensated but at the expense of inducing forward voltage drop. Contrarily in [21], the forward voltage drop is addressed while ignoring the dead-time. There also exist some models which take both the dead-time and forward voltage drop into consideration and both the factors are separately analyzed suggesting the possibility of compensating both the effects and found them independent of each other [22]. The error time and error voltage are mutually convertible as per the average value theory. Since inverter legs operate at disparate legs, the error of each can be independently measured to achieve more accurate compensation. This will involve constant estimation of error as the switching frequency changes [23]. There is however the danger of unexpected current clamp in this method when the compensation suddenly changes while current crosses zero.

In the second category of dead-time compensation techniques, the current direction is not necessary, and a complex model of the adaptive voltage compensation algorithm is implemented to suppress the dead-time effect [24]. A fast and a slow response disturbance observer are employed which makes a distinction between the back EMF and disturbance voltage, leading to voltage error correction [25]. However accurate motor parameters and complex calculations are major downsides to this approach. The effects of parasitic capacitances of power devices are also important in the context of the compensation method [26]. The dead-time causes an error in the modulation voltage. In [27], the dead-time compensation is utilized for the modulation error and the effects of the dead time have been analyzed on three-level inverters. However, the proposed technique on [27] isn't used to eliminate the consequences of dead time effect on power quality.

In this paper, the five parameters of power quality have been improved by a novel DTC technique and the consequences of dead time effect on power quality have been eliminated. A schematic of an industrial microgrid with DTC is given in Figure 1. In this figure, industrial motors make up the majority of the industrial microgrid loads. Introducing DTC in power converters can help in curtailing the non-ideal nature of power electronics. The current research attempts to alleviate the effects of dead time on power converter parameters, for instance, curtailing fundamental voltage, distorting current waveforms, phase angle, and third harmonic as well as torque pulsation. The parametric equations for all the parameters are derived for normal scenarios and dead time compensation scenarios. The model is simulated in MATLAB with two different cases and the results, with and without the proposed model application, are drawn.

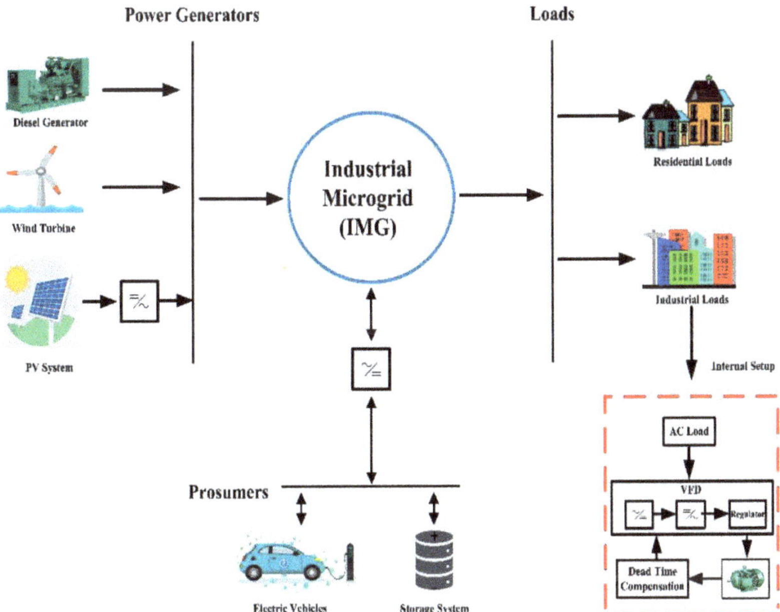

Figure 1. Conceptual view of an Industrial Microgrid.

The article is organized as follows: Section 2 delineates the model structure for power quality enhancement with the DTC technique; Section 3 investigated the proposed DTC method along with the mathematical modeling of this method; Section 4 explains the outcomes of the research work with a discussion on the importance of the research to the field, and Section 5 lists the conclusions derived from the research endeavor.

2. System Modeling

The model of power quality improvement involves the incorporation of high-level pulse width modulation, in this case, IGBT, into the electrical switching system while retaining the switching frequency at 2–15 kHz levels [21]. To cope with the non-ideality in a small delay, called dead time, is inserted in the operation to avoid short circuits. Meanwhile, the dead time causes various parameters deterioration, therefore power quality analysis is carried out which mostly focused on the reduction of the negative consequences of dead time. The proposed method consists of dead time compensation (DTC) strategy for stabilization of 3 phase induction motors in the open-loop system using variable frequency drive (VFD) to govern the speed required for AC motors besides the offsetting of the adverse effects of DTC. A schematic of the proposed model is outlined in Figure 2. It consists of "volt-per-Hz" drive the design of which is delineated in the following. A specially minted control mechanism is applied for maintaining a fixed level of magnetizing current. Additionally, variable stator voltage support is also implemented.

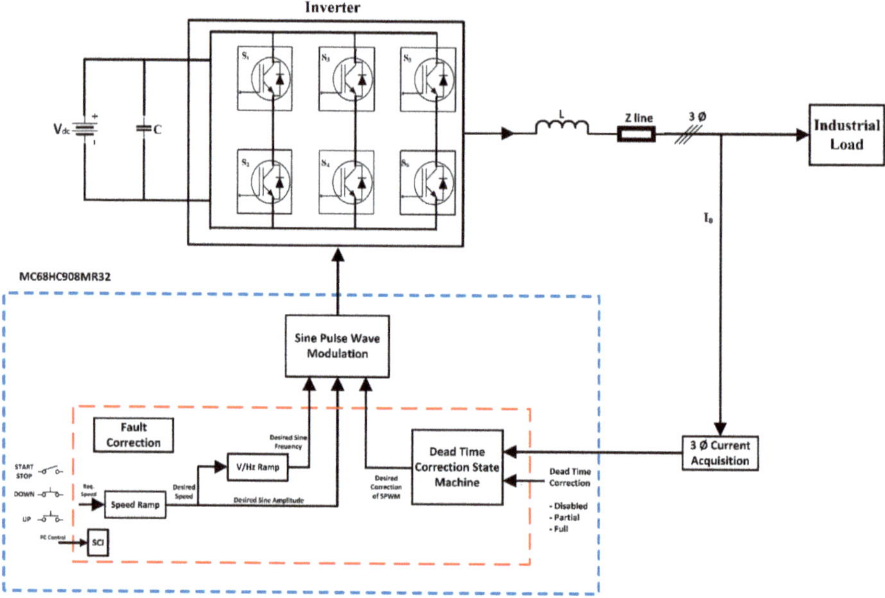

Figure 2. Model structure with dead-time compensation.

Since dead time is directly proportional to PWM signal output; the increase or decrease of one directly decreases or decreases the other resulting in a closer-to-original voltage pulse. Accordingly, the current study employs two methods for correcting dead time induced distortion: full correction methodology which factors in the phase angle magnitude and initiates a novel s/w for enhanced output; and partial correction methodology which helps the PWM on-chip hardware. At least one of these correction methodologies is pertinent to improve the PWM inverter parameters. The increasing parameters of PWM inverter values used in pair register; need to be kept in check with the help of software. The value is dependent on the transistor and is important for the output voltage in DT. The partial correction method essentially builds the basis for polarity detection, which is helpful in the improvement of load current waveform, the magnitude of fundamental voltage, phase angle, third harmonic distortion, and induction motor parameters such as current waveform, torque pulsation, and speed. Albeit many shortcomings are concomitant with this method; the current settles at zero at the point of disturbance.

3. Proposed Dead Time Compensation Technique

In this section, the proposed dead-time compensation (DTC) method is investigated. First, the DTC for the PWM inverter is performed. Then, the modeling of the proposed DTC technique is introduced.

3.1. Dead-Time Compensation for PWM Inverter

The design of a three-phase voltage-based inverter is given in Figure 3 where 'n' and 'o' indicate dc link and induction motor neutral points respectively and IGBTs paired with diodes work as switches. As an example, the effect of dead time and forward voltage drops have been examined in phase A leg. The phase A leg contains four different current paths as demonstrated in Figure 3. The forward voltage drop of IGBT is represented as 'u_{ce}' and that of the anti-diode as 'u_d'. From Figure 4, phase A current i_a is represented by the dashed line in Figure 4. It is established from Figure 4a,b that when current flows from the inverter to load, $i_a > 0$ and from Figure 4c,d it is evident that $i_a < 0$ for opposite current direction reverses.

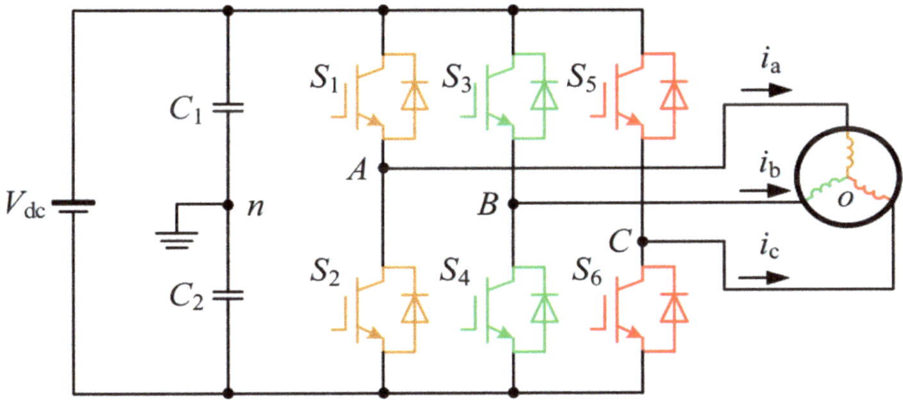

Figure 3. Two-level inverter-motor system [21].

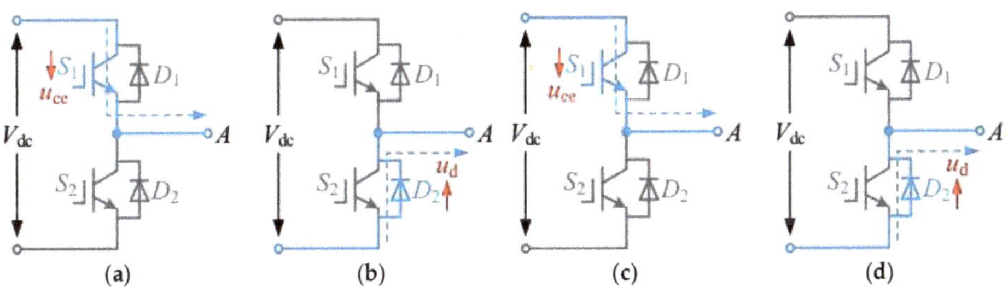

Figure 4. Analysis of different current flow paths of phase A. (**a**) When i_a>0 from D1 (**b**) When i_a>0 from D2 (**c**) When i_a<0 from D1 (**d**) When i_a<0 from D2 [21].

Since the delay in the on-off of IGBTs for this study is extremely small in comparison to the dead time, it is considered negligible. In the beginning, the dead time setting is analyzed. The delays between turn on and off of power electronic devices (IGBTs) are considered negligible in this study as they are very small in comparison to the dead time. One switching cycle of sinusoidal pulse width modulation (SPWM) contains two stages of dead time in which power electronic devices remain in the OFF state. Hence, the load current is forced to pass through anti-parallel diodes D_1 or D_2 (depending on the direction). Current flows through D_2 when i_a is positive during phase A and is interconnected to the negative terminal as evident from Figure 4b. If i_a is negative, current flows through D_1 with phase A connection setting shown in Figure 4c.

Forward voltage drop occurs when the switching devices pass load current. When the current is positive, the voltage output at phase A is marginally less than the respective DC linkage voltage. In such instances, the IGBT S1 passes current from the positive linkage or D2 passes current from the negative linkage. Similarly, for negative i_a, the voltage output at phase A is marginally greater in comparison to the DC linkage voltage. In this case, positive linkage gets current via D1, and negative linkage via S2. The various permutations of currents flow, and voltage waveforms at different instances are given in Table 1.

Table 1. The error between the ideal voltage and the actual voltage.

Time (T)	Ideal O/P Voltage/Actual Voltage	Ideal O/P Voltage	Actual O/P Voltage	Dead-Time Error
$t_0 - t_1$	D_2/D_2	$-V_{dc}/2$	$-V_{dc}/2 - u_d$	u_d
$t_1 - t_2$	S_1/D_2	$V_{dc}/2$	$-V_{dc}/2 - u_d$	$V_{dc} + u_d$
$t_2 - t_3$	S_1/S_1	$V_{dc}/2$	$V_{dc}/2 - u_{ce}$	u_{ce}
$t_3 - 4$	D_2/D_2	$-V_{dc}/2$	$-V_{dc}/2 - u_d$	u_d
$t_4 - t_5$	D_2/D_2	$-V_{dc}/2$	$-V_{dc}/2 - u_d$	u_d
$t_5 - t_6$	S_2/S_2	$-V_{dc}/2$	$-V_{dc}/2 + u_{ce}$	$-u_{ce}$
$t_6 - t_7$	D_1/D_1	$V_{dc}/2$	$V_{dc}/2 + u_d$	$-u_d$
$t_7 - t_8$	D_1/D_1	$V_{dc}/2$	$V_{dc}/2 + u_d$	$-u_d$
$t_8 - t_9$	S_2/D_1	$-V_{dc}/2$	$V_{dc}/2 + u_d$	$-V_{dc} - u_d$
$t_9 - t_{10}$	S_2/S_2	$-V_{dc}/2$	$-V_{dc}/2 + u_{ce}$	$-u_{ce}$

As seen from Table 1 the error voltage is the largest for the time range $t_8 - t_8$ where i_a is negative. These analyses for phase A leg can also be applied to three-phase legs. The gate signals and voltage waveforms of phase A as per the pulse generation rule of SPWM are given in Figure 5. It depicts the gate input and dead time incorporated gate input signals for top and bottom switching devices ($u + g$, $u - g$), voltage output (u_{ideal}), real output voltage incorporating dead time (u_{real}), with both dead time and forward voltage drop (u_{real2}).

Accordingly, the corresponding dead time error is given as the difference between u_{ideal} and u_{ideal2}. Similarly, V_{dc}, T_s, and T_d indicate the DC linkage voltage, switching period, and dead time respectively. It is evident from the left side of Figure 4 that for negative i_a ideal PWM voltage is greater than the actual PWM voltage. In other words, the ideal voltage output will exceed the actual voltage output. The actual o/p voltage will be slightly less when the forward voltage drop is applied. The voltage varies between $\frac{V_{dc}}{2} - u_{ce}$ and $-\frac{V_{dc}}{2} - u_d$ when the load current is applied to positive and negative terminal respectively. It is evident from Figure 5 that the deviation from ideal behavior is a function of dead time. Additionally, there is a dependence on the current direction; when altered from load to positive terminal, the output is $\frac{V_{dc}}{2} + u_d$, while changes to $-\frac{V_{dc}}{2} + u_{ce}$ for the opposite current direction.

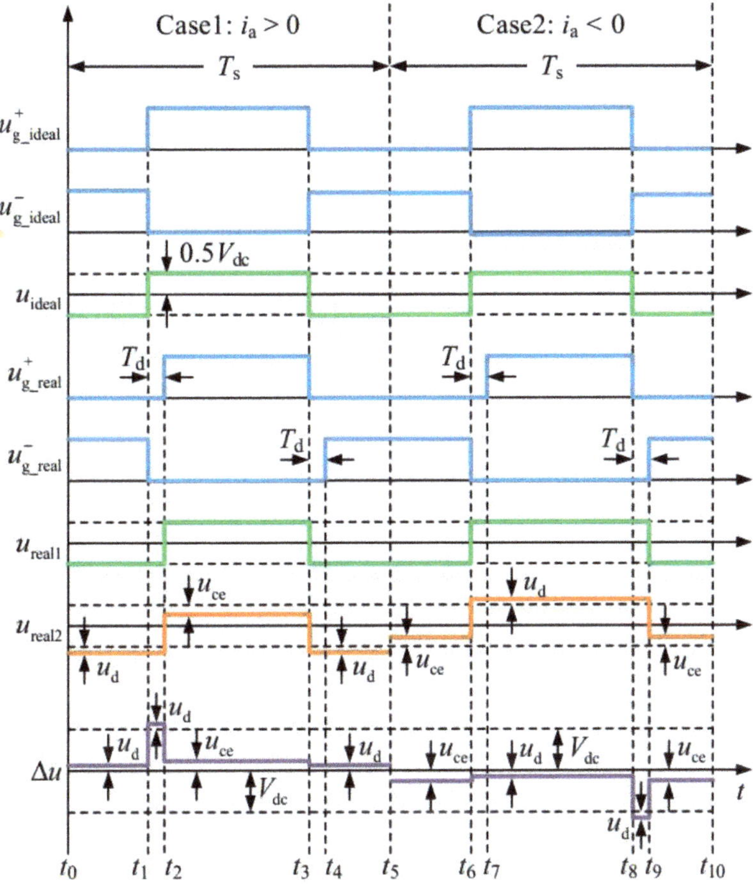

Figure 5. Illustration of voltage waveforms and gate signals [21].

3.2. Modeling of the Proposed Dead Time Compensation

The block diagram for 3-phase idealized PWM inverter is illustrated in Figure 6. The DTC model is schematically presented in Figure 7. The model employs an ideal relay possessing two specifications: memory-less, and nonlinearity. The voltage distortion ε depends on T_d the delay time and carrier signal $V_c(t)$ the slope of the triangular waveform. Take the required signal $V_i(t)$ is gradually varying as compared to the high-frequency carrier signal C. The ratio ε/T_d is equal to $2V_c/\left(\frac{T_c}{2}\right)$ the down-slope, the triangular carrier signal $V_c(t)$ and therefore we have $\varepsilon = 2V_cT_d/\left(\frac{T_c}{2}\right) = 4f_cV_cT_d$. Where ε represents as voltage distortion, T_d is a time delay and f_c represents the frequency of the carrier signal.

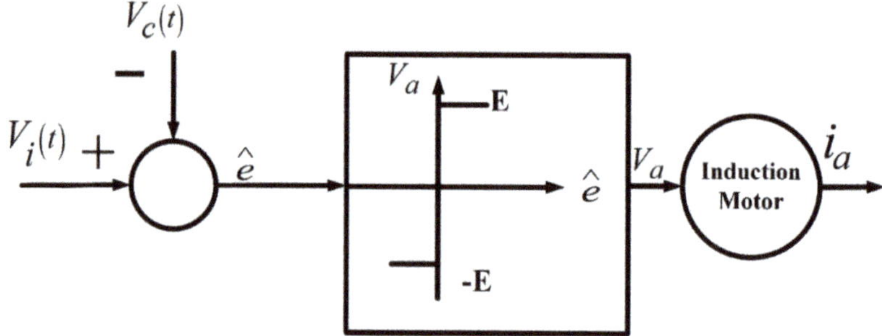

Figure 6. Block diagram for 3-phase idealized pulse width modulation (PWM) inverter.

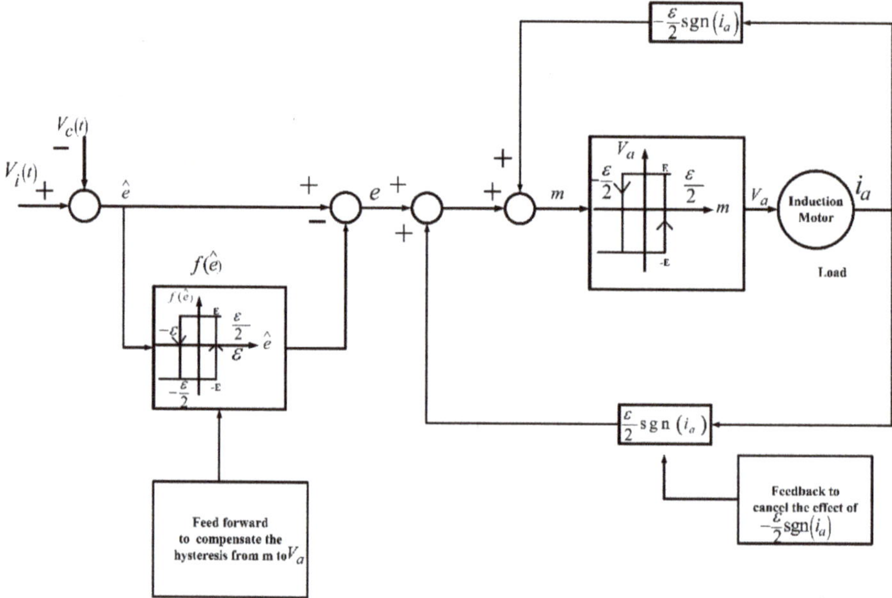

Figure 7. Proposed model for the (DTC) method.

The desired results can be obtained by setting the deviation 'ε' to 0 or dead time 'T_d' to 0. Since the inherent objective is to compensate for the dead time, the following derivation further explores the ideal PWM inverter [27].

$$T_{err} = T_{off} - T_{on} - T_d + 2T_{com} \tag{1}$$

where 'T_{com}' indicates the compensation time when '$i_a < 0$'.

For phase A:

$$T_{err\_a} = Sign(i_a)T_{ma} \tag{2}$$

where

$$T_{ma} = T_{off} - T_{on} - T_d + 2T_{com} \tag{3}$$

$$Sign(i_a) = \begin{cases} 1 & (i_a > 0) \\ 1 & (i_a < 0) \end{cases} \tag{4}$$

When voltage U_a is positive for phase A, the switching cycle time duration is T_a, and for S_1, it is T_a^*. For negative U_a, the switching cycle time duration is T_a for phase A, and for S_4 it is T_a^*.

Correspondingly the relation between effective time (T_a) and commanded time (T_a^*) comes out to be [21]:

$$T_a = T_a^* + Sign(i_a)T_{ma} \tag{5}$$

Similarly, for Phase B and C the same quantities are given as:

$$T_b = T_b^* + Sign(i_b)T_{mb} \tag{6}$$

$$T_c = T_c^* + Sign(i_c)T_{mc} \tag{7}$$

where IGBTs S_3 and S_1 remain turned on and off respectively for (1) to (7). However, at the neutral point when fundamental voltage is balanced:

$$V_{dc1} = V_{dc2} = \frac{V_{dc2}}{2}$$

when $i_a > 0$:

$$V_{ao} = \frac{V_{dc}}{2} - 2V_{ce} \; (when \; S_a = 1) \tag{8}$$

$$V_{ao} = -V_d - V_{ce} \; (when \; S_a = 0) \tag{9}$$

$$V_{ao} = -\frac{V_{dc}}{2} + 2V_{ce} \; (when \; S_a = -1) \tag{10}$$

when $i_a < 0$;

$$V_{ao} = \frac{V_{dc}}{2} + 2V_d \; (when \; S_a = 1) \tag{11}$$

$$V_{ao} = V_d + V_{ce} \; (when \; S_a = 0) \tag{12}$$

$$V_{ao} = -\frac{V_{dc}}{2} + 2V_c \; (when \; S_a = -1) \tag{13}$$

Supposing no change in the direction of current, (8)–(13) gives:

$$V_{ao} = S_a\left(\frac{1}{2}V_{dc} + V_d - V_{ce}\right) - Sign(i_a)(V_{ce} + V_d) \tag{14}$$

When voltage drop increases with respect to current:

$$V_{ce} = V_{ceo} + r_{ce}|i_a| \tag{15}$$

$$V_d = V_{do} + r_{rd}|i_a| \tag{16}$$

Combining (15) and (16) with (14) gives:

$$V_{ao} = S_a\left(\frac{1}{2}V_{dc} + V_d - V_{ce}\right) - Sign(i_a)(V_{ce} + V_d) \tag{17}$$

As per volt-second balance theorem:

$$S_a = \left[\frac{T_a^* + T_{ma}Sign(i_a)}{T_s}\right]Sign\left(U_{a\_ref}\right) \tag{18}$$

$$V_a = \left[\frac{T_a^* + T_{ma}Sign(i_a)}{T_s}\right]\left(\frac{1}{2}V_{dc} + V_d - V_{ce}\right)Sign\left(U_{a\_ref}\right) - (V_{ceo} + V_{do})Sign(i_a) - (r_{ce} + r_d)i_a \tag{19}$$

Now for phase b and c:

$$V_b = \left[\frac{T_b^* + T_{mb}Sign(i_b)}{T_s}\right]\left(\frac{1}{2}V_{dc} + V_d - V_{ce}\right)Sign\left(U_{b\_ref}\right) - (V_{ceo} + V_{do})Sign(i_b) - (r_{ce} + r_d)i_b \tag{20}$$

$$V_c = \left[\frac{T_c^* + T_{ma}Sign(i_c)}{T_s}\right]\left(\frac{1}{2}V_{dc} + V_d - V_{ce}\right)Sign\left(U_{c\_ref}\right) - (V_{ceo} + V_{do})Sign(i_c) - (r_{ce} + r_d)i_c \qquad (21)$$

Balanced load for three-phase loads is indicated as:

$$V_a + V_b + V_c = 0 \qquad (22)$$

$$i_a + i_b + i_c = 0 \qquad (23)$$

$$\begin{cases} V_a = V_{ao} + V_o \\ V_b = V_{bo} + V_o \\ V_c = V_{co} + V_o \end{cases} \qquad (24)$$

Hence, the schematic illustration of the PWM inverter shown in Figure 3 is thus transformed into the final DTC model schematically represented in Figure 8.

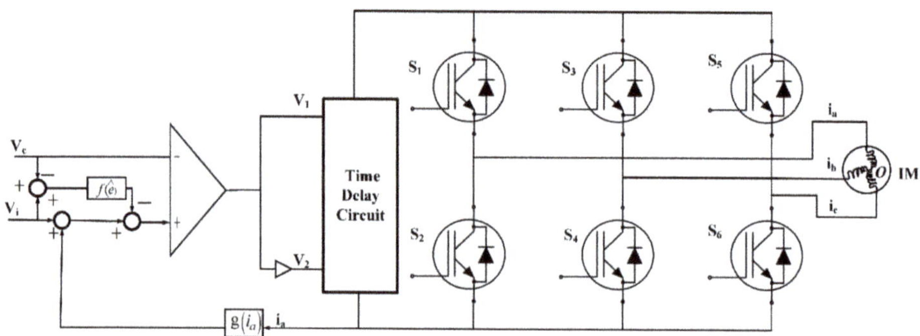

Figure 8. Three-phase SPWM inverter with the proposed DTC method.

Subsequently, the control block of the upper section $\left(-\left(\frac{\varepsilon}{2}\right)sign(i_a)\right)$ is canceled out with a feedback block $\left(g(i_a) = \left(\frac{\varepsilon}{2}\right)sign(i_a)a\right)$. The Feed-forward $f(\hat{e})$ method is employed for dealing with hysteresis. The hysteresis compensation block as given in Figure 7 is utilized. To achieve the characteristics of an ideal relay, transfer features such as m, V_a are used. Figure 7 also gives a representation of the dead time feedback blocks $g(i_a)$, and feed-forward block $f(\hat{e})$ of the SPWM inverter model. Owing to the inherent phase lag the feed-forward compensation cannot be ignored. At this stage, the DTC technique can be applied to the 3-phase PWM inverter, an example of which has been demonstrated in Figure 3.

The compensation of dead time blocks $g(i_a)$ (feedback) and $f(e)$ (feedforward) of the 3 phase SPWM inverter model is presented in Figure 7 schematically. Due to an inherent phase lag, the $f(\hat{e})$ (feed-forward compensation) cannot be ignored. Just now at this stage, the DTC technique is ready to be applied on 3-phase PWM inverter for practical implementation which is presented in Figure 3.

Furthermore, before the dead time compensation method the equation has the following shape:

$$e = [V_i(t)] - V_c(t) \qquad (25)$$

After converting the deviation, ε, to phase voltage, V_a; adjusting T_d and ε equal to zero; eliminating the factor $-\left(\frac{\varepsilon}{2}\right)sign(i_a)$ through feedback factor $g(i_a) = \left(\frac{\varepsilon}{2}\right)sign(i_a)$ application; and applying for the compensation through feedforward for dealing with the inherent phase angle.

It is pertinent that the voltage control signal $V_i(t)$ in (25) is equal to $[V_i(t) - f(\hat{e}) + g(i_a)]$ and Equation (25) becomes:

$$e = [V_i(t) - f(\hat{e}) + g(i_a)] - V_c(t) \qquad (26)$$

where $\hat{e} = V_i - V_c$ and $f(e)$ and $g(i_a)$ are nonlinear functions.

4. Results and Discussion

Following the mathematical modeling, the proposed dead-time compensation model was simulated using MATLAB/Simulink for validation of the method. The V/f strategy was employed for system control. Since the compensation strategy only depends on the characteristics of the power devices, a three-phase Y-connected symmetrical RL load was deployed at the output terminal of the inverter. The key parameters used in the simulation are listed in Table 2 while the important characteristics of the industrial motor are given in Table 3. The simulation time is kept twice the fundamental period to avoid imprecise results and surplus data; otherwise, the simulation may stop due to computer memory exhaustion.

Table 2. Relevant parametric values of the Simulink Model.

No	Parameters	Input Values
1	Reference Signal f_r	50 Hz
2	Carrier Signal f_c	5k Hz
3	Amplitude modulation Index	0.8
4	DC Voltage	700 V
5	Time Delay T_d	10 µs
6	V_c	1
7	V_r	0.8
8	Load Resistance	12.6 Ω
9	Load inductor	40 mH

Table 3. Characteristics of induction motor (industrial load).

No	Parameters	Rating
1	Nominal power of IM	5.4 HP
2	The nominal voltage of induction motor	400 V
3	Nominal frequency of induction motor	50 Hz
4	Speed	1430 rpm
5	Power factor	0.8
6	Rated torque	10 Nm

Dead-time distortion correction algorithms are a useful tool for adjusting PWM relevant to the actual polarity of phase current. The PWM control signal is extended by the addition of dead time, to match the actual pulse with the desired values, when the voltage pulse is shortened. Contrarily, for prolonged voltage pulse by dead time, the PWM signal is reduced by an equivalent time, leading to a match between the actual and desired voltage pulse. Resultantly an actual voltage signal equal to the desired signal is achieved, along with a sinusoidal phase current.

4.1. Impact of Dead Time on Load Current Waveform

Without the dead time compensation and the proposed dead time compensation method, when the fundamental frequency f_1 is 5 Hz, the load current waveform is substantially improved. The amplitude of the current waveform is increased and distortion is reduced significantly. The current waveform is almost the same as the ideal current waveform. The provision for the mandatory delay in switching signals in IGBTs to accommodate dead time can induce undesirable sub-harmonics, subsequently causing deviation in load current as illustrated in Figure 9. The proposed DTC compensates the distortions to make the signal more sinusoidal as shown in Figure 10.

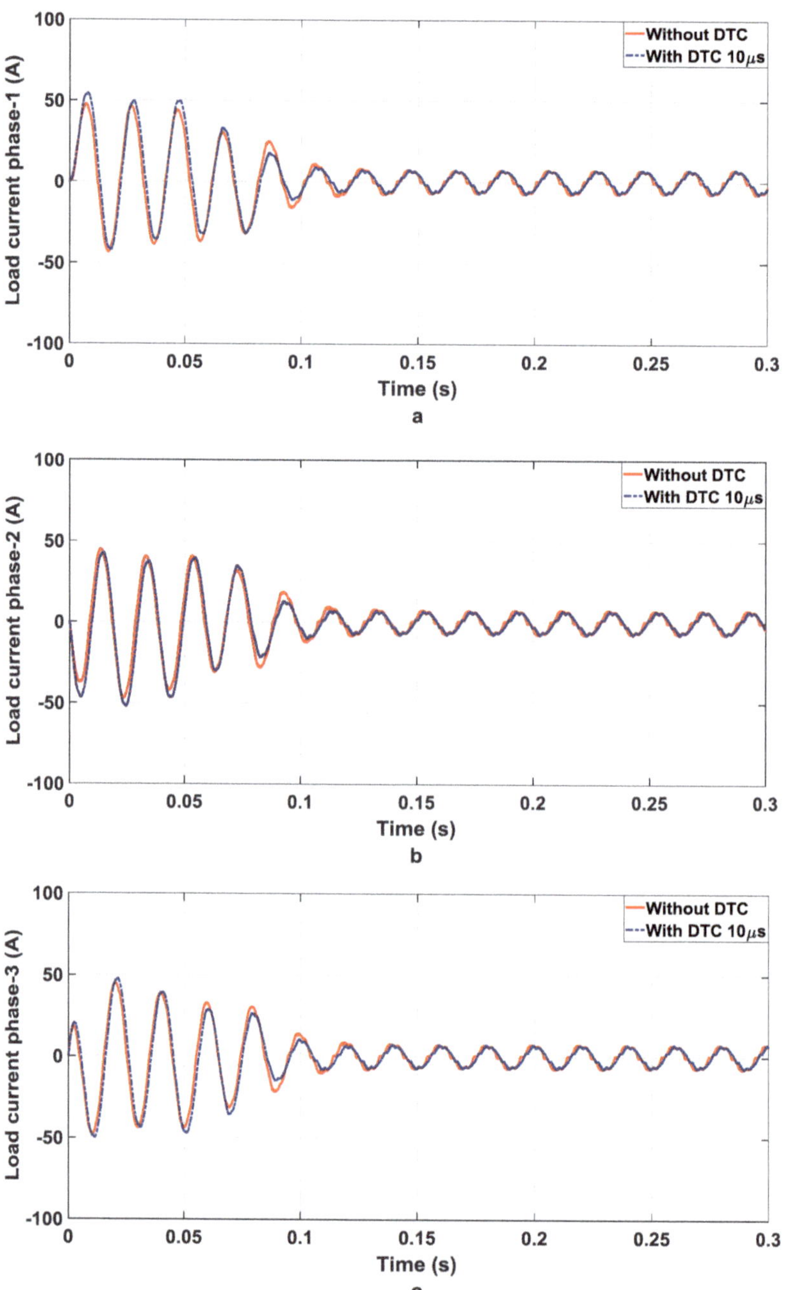

Figure 9. Load current with/without DTC (**a**) phase 1 (**b**) phase 2 (**c**) phase 3.

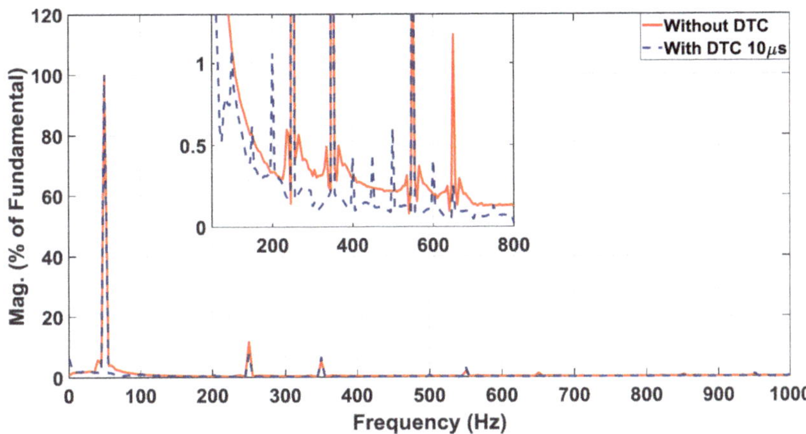

Figure 10. Calculation of total harmonic distortion (THD) without and with DTC.

4.2. Total Harmonics and Individual Harmonics Distortions Calculation by FFT Analysis

The third harmonic distortion is the main problem due to the non-ideal characteristics of power converters. By nature, 'each regularly distorted waveform may be defined as a number of pure sine waves in which the frequency of each sinusoid is an integer multiple of the fundamental frequency of the distorted wave. The sum of the sinusoids is referred to as the "Fourier series." In recent years, they have also concentrated on the harmonic distortion of the power field.

Dead-time is unavoidable in inverter circuitry as it prevents short-circuiting. However, it comes with the side effect of total harmonic distortion, thus necessitating DTC. The proposed DTC can alleviate the side effects. As can be seen from Figure 10, the third harmonic distortion is 16.26% without DTC. However, after the application of the novel DTC, the distortion is mitigated by 3.77% to 12.49%, as presented in Table 4. This improvement of almost 4% will be instrumental for the health of the motors operating in industrial load.

Table 4. FFT analysis of the load current.

	Without DTC	With DTC 10 µs
Fundamental Frequency (50 Hz)	7.437	6.784
Total harmonics distortion THD (%)	16.26%	12.49%

Individual harmonic distortion (IHD) represents the relation between the root mean square (RMS) value of the fundamental (RMS) value of the individual harmonics in Equation (29) [28]:

$$IHD_n = I_n / I_1 \tag{29}$$

For third harmonic, n is represented by 3. From Figure 10, the RMS of the fundamental current is equal to 100. Also, the RMS of the third harmonic current without DTC is 11.615 A and with DTC is 7.475 A. Therefore, $IHD_3 = 11.615 / 100 * 100 = 11.615$ without DTC and $IHD_3 = 7.475$ with DTC.

4.3. Improvement in Fundamental Voltage Magnitude and Phase Angle

Dead-time induces certain drawbacks in the power electronic circuitry such as a decrease in fundamental voltage and distortion of other parameters. These effects can be effectively coped with through the incorporation of the proposed DTC. The fundamental voltage magnitude can be restored, and the harmonics minimized. Figure 11a for dead time 10 µs presents the fundamental voltage magnitude in the absence of DTC and presents the same parameter after DTC, respectively. It is evident that the fundamental voltage has significantly improved as a result of the application of the proposed DTC technique.

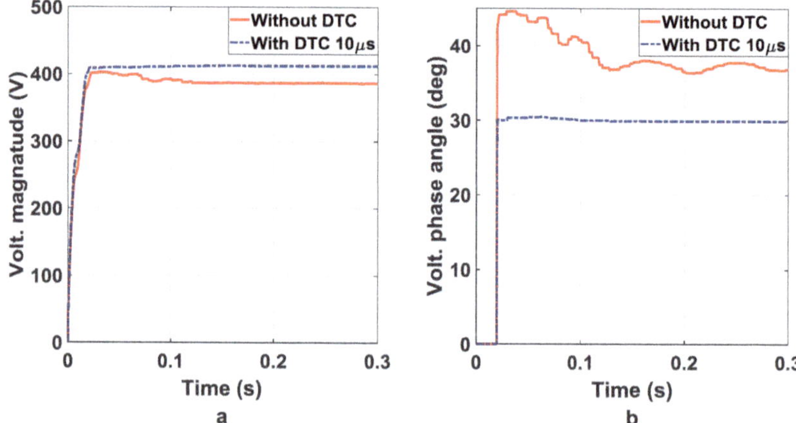

Figure 11. Without/with 10 µs DTC (**a**) fundamental voltage (V_a) magnitude (**b**) fundamental voltage (V_a) phase angle.

Phase angle distortion is also a downside of dead time. However, an efficient DTC strategy can handle this drawback to a certain degree. The proposed DTC model can achieve significant improvements in this domain as well. Figure 11b for dead time 10 µs shows the phase angles distortion pre and posts using the DTC technique respectively. It is vividly evident from these figures that the phase angle distortion has been significantly reduced through the application of the proposed DTC technique.

4.4. Improvement in Power Quality Parameters of Induction Motor

The induction motors are major energy-using equipment, any issues with their smooth operation are extremely significant. The unavoidable delay in signal switching can cause sub-harmonics leading to waveform distortion in the current signal. Additionally, it can also lead to pulsation in torque, and reduction in the rotational speed of the motor, manifesting in heat dissipation from the motors. These distortions and the concomitant damages can be significantly reduced by employing an efficient DTC technique. The DTC technique can restore the current to one looking more like the sinusoidal curve which entails the remedy of the aforementioned drawbacks in motor performance. DTC insertion in the inverter circuitry has been shown to mitigate the harmonic distortion and load torque pulsation. Furthermore, DTC can lead to practical improvements in motor performance such as smooth operation, limited torque ripples, low noise, and enhance efficiency in operation due to lower harmonic losses.

The proposed DTC technique has been applied to the inverter circuit with the harmonic distortion mentioned above. Figure 12a–c show the improvement in the current waveform distortion, motor speed, and torque pulsation respectively. In Figure 12a–c, using DTC application shows better performance in comparison with using DTC application for the motor speed and the torque pulsation, respectively. The left side of these figures represents these parameters before the application of DTC, while the right sides represent the post DTC parameters' behavior.

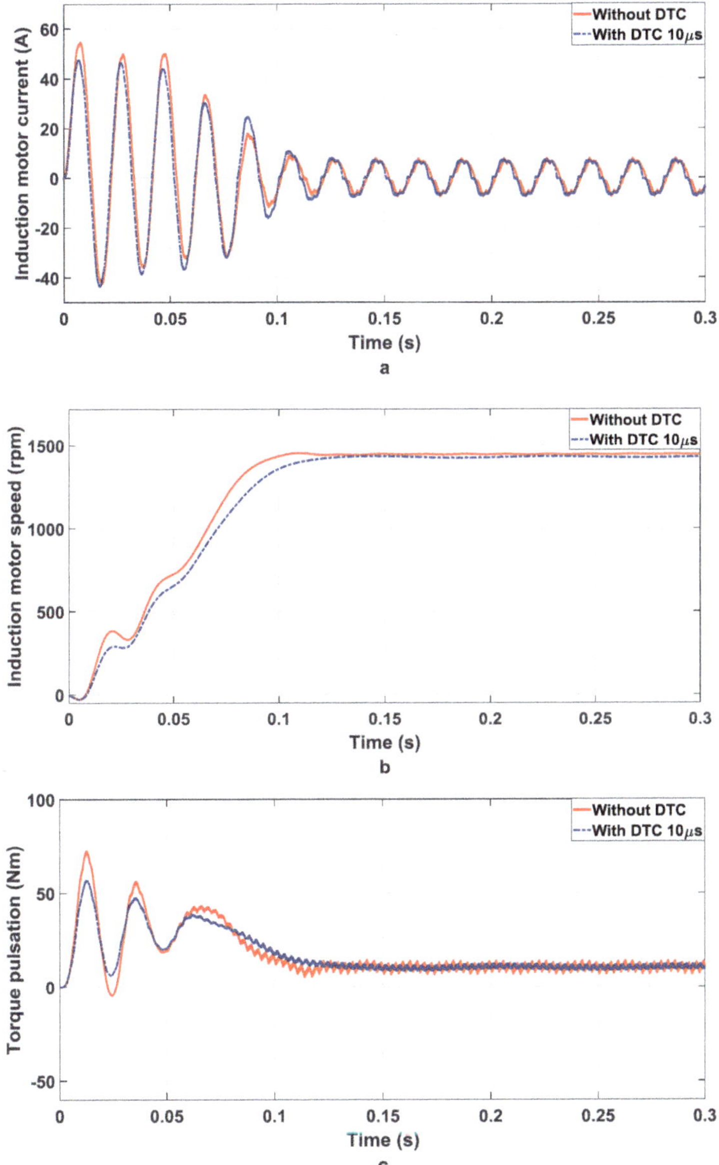

Figure 12. Without/with 10 μs DTC (**a**) induction motor current (**b**) speed of induction motor (**c**) Torque pulsation.

4.5. Case 2: Dead-Time 1 μsec and Switching Frequency 1 kHz

In case 2, the studies show that the current magnitude of the harmonics is the same, while the distortion factors are different due to time increasing as shown in Figure 13. If we take 1 micro-second, as shown in Figure 13, the third harmonic distortion without DTC is 16.26 percent. Moreover, after the implementation of the proposed DTC, method the distortion is reduced from 18.36% to 10.20%, as seen in Table 5. For the quality of power in industrial loads, an average increase of almost 8% is important.

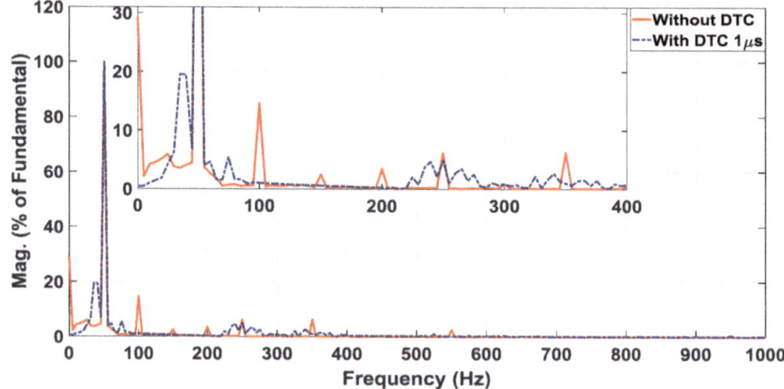

Figure 13. Calculation of total harmonic distortion (THD) without and with DTC.

Table 5. FFT analysis of the load current.

	Without DTC	With DTC 1 µs
Fundamental Frequency (50 Hz)	8.227	6.714
Total harmonics distortion THD (%)	19.20%	16.16%

From Figure 13, the RMS of the fundamental current is equal to 100. Also, the RMS of the third harmonic current without DTC is 15.707 A and with DTC is 10.295 A. Therefore, $IHD_3 = 29.2759/100 * 100 = 29.2759\%$ without DTC and $IHD_3 = 14.645\%$ with DTC.

Similarly, the fundamental voltage magnitude may be recovered in case 2 and the harmonics reduced as shown in Figure 14a. Figure 14a explicitly indicates the voltage magnitude in the presence of the DTC and also displays the same parameter before the DTC when time is 1 microsecond. It is clear that the voltage magnitude has dramatically changed as a result of the implementation of the new DTC methodology.

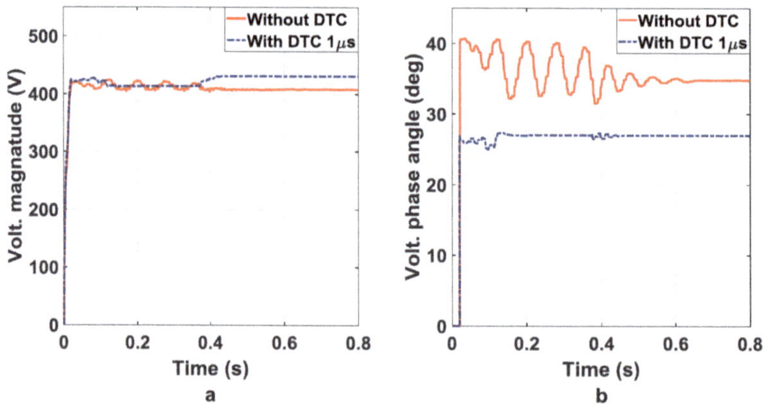

Figure 14. With/without DTC (**a**) fundamental voltage (V_a) magnitude (**b**) fundamental voltage (V_a) phase angle.

In fact, the distortion of the phase angle in case 2 is also increased due to dead-time effects. However, if we take 1 micro-second, an effective DTC strategy will deal with this problem to some large extent. Figure 14b demonstrates the before and after the distorted phase angles of voltage magnitude by using the DTC methodology, respectively. It is clear from this figure that the distortion of the phase angle was greatly decreased by the implementation of the new DTC strategy.

In case 2 while the dead time is changed from 10 to 1 micro-seconds, the DTC incorporation in the inverter circuit is shown to reduce the total harmonic distortions and also torque pulsation as shown in Figure 15a. In addition, DTC will lead to realistic changes in an induction motor performance, along with the smooth operation, low noise, minimal torque ripples, and increased operating efficiency due to the lower harmonic losses. Figure 15a–c demonstrate the increase in current waveform quality, induction motor speed, and also torque pulsation, collectively.

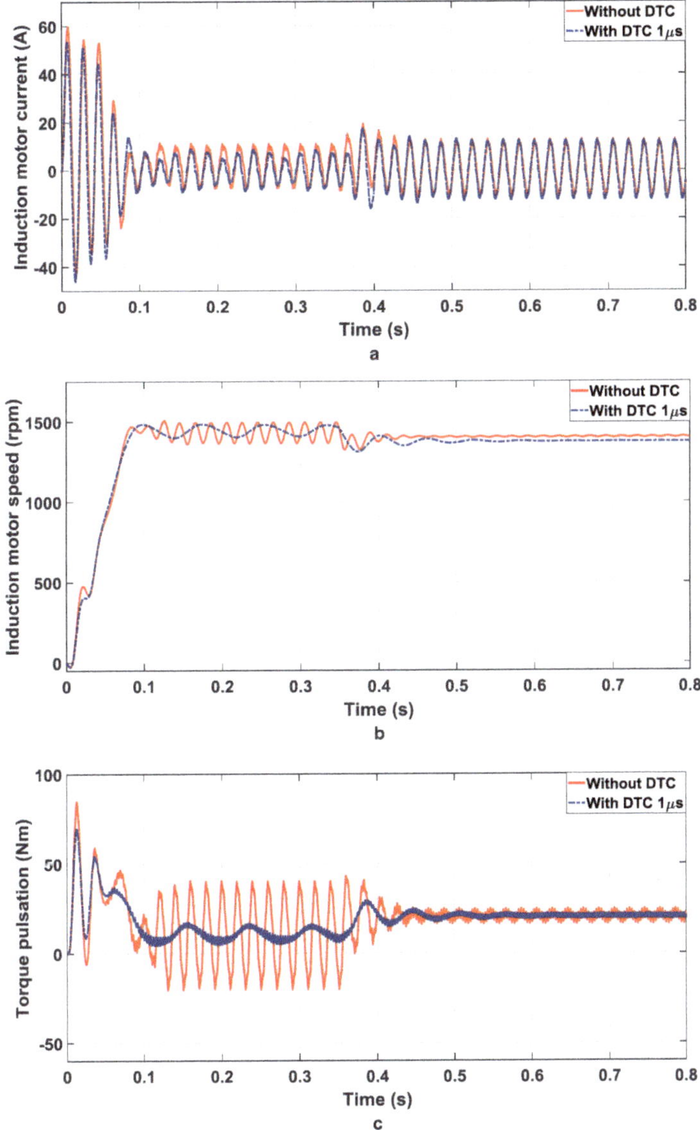

Figure 15. Without/with 1 µs DTC (**a**) induction motor current (**b**) speed of induction motor (**c**) Torque pulsation.

5. Conclusions

The research endeavor successfully models and implements a novel dead time technique rooted in dead time compensation for enhancing the power quality parameters. The overall efficiency improvement of power converters has been achieved from cumulative improvements in several power quality parameters such as sinusoidal load current, phase angle, fundamental voltage magnitude, harmonic distortion. Further, the mitigation in a motor operation like torque pulsation smoothening, current waveform restoration, and speed enhancement are also enhanced. The parametric equations for all the parameters are derived for normal scenarios and dead time compensation scenarios. The model is simulated in MATLAB with two different cases and the results, with and without the proposed model application, are drawn. Finally, two separate case studies of the performance relative to the pre-model and major changes in all dimensions of the power quality parameters and the output of the power converters deficiency are found by the implementation of the dead time compensation technique. The proposed DTC results in significant improvement in the following parameters as exhibited in Figures 9–15:

(a) Sinusoidal load current waveform has been achieved by applying dead time compensation, leading to the removal of distortions accruing from harmonics

(b) Fundamental voltage magnitude has been significantly restored through the proposed DTC implementation.

(c) The phase angle has been improved through the proposed DTC strategy. Additionally, the third harmonic distortion has been significantly alleviated 3.77%, through the proposed DTC strategy in Case 1 and 3.04% in Case 2.

(d) Significant improvements have been achieved in the induction motor parameters post the DTC strategy application. Parameters like torque pulsation, speed, and THD which cause hindrance in smooth motor operation have been demonstrably improved through the novel technique.

Author Contributions: Conceptualization, S.I.; Data curation, M.A.A., M.A., and S.A.A.R.; Formal analysis, M.U.J., M.A.A., M.A., and N.A.S.; Funding acquisition, A.X.; Investigation, H.U.R., S.S., S.A.A.R., and N.A.S.; Methodology, S.I. and A.X.; Project administration, A.X.; Resources, A.X.; Software, S.I.; Supervision, A.X.; Validation, M.U.J., and M.A.A.; Visualization, H.U.R., and S.S.; Writing—original draft, S.I.; Writing—review & editing, A.X., M.U.J., and M.A.A. All authors have read and agreed to the published version of the manuscript.

Funding: This work was supported by the 2016 national key R & D program of China to support low-carbon Winter Olympics of integrated smart grid demonstration project (2016YFB0900500) and Beijing Natural Science Foundation (3182037). The Fundamental Research Funds for the Central Universities under Grant 2019QN041.

Conflicts of Interest: The authors declare no conflict of interest.

References

1. Nejabatkhah, F.; Li, Y.W.; Tian, H. Power quality control of smart hybrid AC/DC microgrids: An overview. *IEEE Access* **2019**, *7*, 52295–52318. [CrossRef]

2. Aziz, T.; Ahmed, M.; Nahid-Al-Masood. Investigation of harmonic distortions in photovoltaic integrated industrial microgrid. *J. Renew. Sustain. Energy* **2018**, *10*, 053507. [CrossRef]

3. Abdelbaky, M.A.; Liu, X.; Jiang, D. Design and implementation of partial offline fuzzy model-predictive pitch controller for large-scale wind-turbines. *Renew. Energy* **2020**, *145*, 981–996. [CrossRef]

4. Kong, X.; Ma, L.; Liu, X.; Abdelbaky, M.A.; Wu, Q. Wind Turbine Control Using Nonlinear Economic Model Predictive Control over All Operating Regions. *Energies* **2020**, *13*, 184. [CrossRef]

5. Iqbal, S.; Xin, A.; Jan, M.U.; Salman, S.; Zaki, A.u.M.; Rehman, H.U.; Shinwari, M.F.; Abdelbaky, M.A. V2G Strategy for Primary Frequency Control of an Industrial Microgrid Considering the Charging Station Operator. *Electronics* **2020**, *9*, 549. [CrossRef]

6. Gandoman, F.H.; Ahmadi, A.; Sharaf, A.M.; Siano, P.; Pou, J.; Hredzak, B.; Agelidis, V.G. Review of FACTS technologies and applications for power quality in smart grids with renewable energy systems. *Renew. Sustain. Energy Rev.* **2018**, *82*, 502–514. [CrossRef]

7. Belov, V.; Butkina, A.; Bolschikov, F.; Leisner, P.; Belov, I. Power quality and EMC solutions in micro grids with energy-trading capability. In Proceedings of the 2014 International Symposium on Electromagnetic Compatibility, Gothenburg, Sweden, 1–4 September 2014; pp. 1203–1208.

8. Standard IEC 61000-2-2. *Electromagnetic Compatibility (EMC)-Part 2-2: Environment-Compatibility Levels for Low-Frequency Conducted Disturbances and Signalling in Public Low-Voltage Power Supply Systems*; IEC: Geneva, Switzerland, 2002.

9. Dong, H.; Yuan, S.; Han, Z.; Ding, X.; Ma, S.; Han, X. A comprehensive strategy for power quality improvement of multi-inverter-based microgrid with mixed loads. *IEEE Access* **2018**, *6*, 30903–30916. [CrossRef]

10. Elbasuony, G.S.; Aleem, S.H.A.; Ibrahim, A.M.; Sharaf, A.M. A unified index for power quality evaluation in distributed generation systems. *Energy* **2018**, *149*, 607–622. [CrossRef]

11. Martinenas, S.; Knezović, K.; Marinelli, M. Management of power quality issues in low voltage networks using electric vehicles: Experimental validation. *IEEE Trans. Power Deliv.* **2016**, *32*, 971–979. [CrossRef]

12. Szczesniak, P. Challenges and design requirements for industrial applications of AC/AC power converters without DC-link. *Energies* **2019**, *12*, 1581. [CrossRef]

13. Guha, A.; Narayanan, G. Small-signal stability analysis of an open-loop induction motor drive including the effect of inverter deadtime. *IEEE Trans. Ind. Appl.* **2015**, *52*, 242–253. [CrossRef]

14. Guha, A.; Narayanan, G. Impact of undercompensation and overcompensation of dead-time effect on small-signal stability of induction motor drive. *IEEE Trans. Ind. Appl.* **2018**, *54*, 6027–6041. [CrossRef]

15. Lin, Y.-K.; Lai, Y.-S. Dead-time elimination of PWM-controlled inverter/converter without separate power sources for current polarity detection circuit. *IEEE Trans. Ind. Electron.* **2009**, *56*, 2121–2127.

16. Dafang, W.; Bowen, Y.; Cheng, Z.; Chuanwei, Z. A feedback-type phase voltage compensation strategy based on phase current reconstruction for ACIM drives. *IEEE Trans. Power Electron.* **2013**, *29*, 5031–5043. [CrossRef]

17. Shen, Z.; Jiang, D. Dead-time effect compensation method based on current ripple prediction for voltage-source inverters. *IEEE Trans. Power Electron.* **2018**, *34*, 971–983. [CrossRef]

18. Lim, J.-W.; Bu, H.; Cho, Y. Novel Dead-Time Compensation Strategy for Wide Current Range in a Three-Phase Inverter. *Electronics* **2019**, *8*, 92. [CrossRef]

19. Iqbal, S.; Xin, A.; Jan, M.U.; Rehman, H.; Salman, S.; Rizvi, S.A.A. Improvement in the Efficiency of Inverter Involved in Microgrid. In Proceedings of the 2018 2nd IEEE Conference on Energy Internet and Energy System Integration (EI2), Beijing, China, 20–22 October 2018; pp. 1–5.

20. Huang, Z.-L.; Jiang, Z.; Li, Z.-H. Dead-time Compensation Strategy of Three Level Inverter Based on FFT. *DEStech Trans. Eng. Technol. Res.* **2019**. [CrossRef]

21. Zhao, L.; Song, W.; Feng, J. A Compensation Method of Dead-Time and Forward Voltage Drop for Inverter Operating at Low Frequency. *J. Electr. Eng. Technol.* **2019**, *14*, 781–794. [CrossRef]

22. Ji, Y.; Yang, Y.; Zhou, J.; Ding, H.; Guo, X.; Padmanaban, S. Control strategies of mitigating dead-time effect on power converters: An overview. *Electronics* **2019**, *8*, 196. [CrossRef]

23. Mousavi, S.Y.M.; Jalilian, A.; Savaghebi, M. Voltage unbalance compensation by a grid connected inverter using virtual impedance and admittance control loops. In Proceedings of the 2018 9th Annual Power Electronics, Drives Systems and Technologies Conference (PEDSTC), Tehran, Iran, 13–15 February 2018; pp. 174–179.

24. Wang, D.; Peng, F.; Ye, J.; Yang, Y.; Emadi, A. Dead-time effect analysis of a three-phase dual-active bridge DC/DC converter. *IET Power Electron.* **2017**, *11*, 984–994. [CrossRef]

25. Kim, D.-Y.; Won, I.-K.; Lee, J.-H.; Won, C.-Y. Efficiency improvement of synchronous boost converter with dead time control for fuel cell-battery hybrid system. *J. Electr. Eng. Technol.* **2017**, *12*, 1891–1901.

26. Mora, A.; Juliet, J.; Santander, A.; Lezana, P. Dead-time and semiconductor voltage drop compensation for cascaded H-bridge converters. *IEEE Trans. Ind. Electron.* **2016**, *63*, 7833–7842. [CrossRef]

27. Guo, J.; Gong, X.; Zhao, F.; Wen, X. A novel dead-time compensation strategy of three-level inverter. In Proceedings of the 2014 IEEE Conference and Expo Transportation Electrification Asia-Pacific (ITEC Asia-Pacific), Beijing, China, 31 August–3 September 2014; pp. 1–5.

28. Sankaran, C. *Power Quality*; CRC press: Boca Raton, FL, USA, 2017.

Article

A Sliding Surface for Controlling a Semi-Bridgeless Boost Converter with Power Factor Correction and Adaptive Hysteresis Band

José Robinson Ortiz-Castrillón [1], Gabriel Eduardo Mejía-Ruiz [2], Nicolás Muñoz-Galeano [1], Jesús María López-Lezama [1,*] and Juan Bernardo Cano-Quintero [1]

[1] Research Group for Efficient Energy Management (GIMEL), Universidad de Antioquia (UdeA), Medellín 050010, Colombia; jrobinson.ortiz@udea.edu.co (J.R.O.-C.); nicolas.munoz@udea.edu.co (N.M.-G.); bernardo.cano@udea.edu.co (J.B.C.-Q.)

[2] Research Group I2, Corporación Universitaria Minuto de Dios—UNIMINUTO, Bello 760101, Colombia; gabriel.mejia@uniminuto.edu

* Correspondence: jmaria.lopez@udea.edu.co

Citation: Ortiz-Castrillón, J.R.; Mejía-Ruiz, G.E.; Muñoz-Galeano, N.; López-Lezama, J.M.; Cano-Quintero, J.B. A Sliding Surface for Controlling a Semi-Bridgeless Boost Converter with Power Factor Correction and Adaptive Hysteresis Band. *Appl. Sci.* **2021**, *11*, 1873. https://doi.org/10.3390/app 11041873

Academic Editors: Oswaldo Lopez Santos and Edris Pouresmaeil

Received: 19 January 2021
Accepted: 16 February 2021
Published: 20 February 2021

Abstract: This paper proposes a new sliding surface for controlling a Semi-Bridgeless Boost Converter (SBBC) which simultaneously performs Power Factor Correction (PFC) and DC bus regulation. The proposed sliding surface is composed of three terms: First, a normalized DC voltage error term controls the DC bus and rejects DC voltage disturbances. In this case, the normalization was performed for increasing system robustness during start-up and large disturbances. Second, an AC current error term implements a PFC scheme and guarantees fast current stabilization during disturbances. Third, an integral of the AC current error term increases stability of the overall system. In addition, an Adaptive Hysteresis Band (AHB) is implemented for keeping the switching frequency constant and reducing the distortion in zero crossings. Previous papers usually include the first and/or the second terms of the proposed sliding surface, and none consider the AHB. To be best of the author's knowledge, the proposed Sliding Mode Control (SMC) is the first control strategy for SBBCs that does not require a cascade PI or a hybrid PI-Sliding Mode Control (PI-SMC) for simultaneously controlling AC voltage and DC current, which gives the best dynamic behavior removing DC overvoltages and responding fast to DC voltage changes or DC load current perturbations. Several simulations were carried out to compare the performance of the proposed surface with a cascade PI control, a hybrid PI-SMC and the proposed SMC. Furthermore, a stability analysis of the proposed surface in start-up and under large perturbations was performed. Experimental results for PI-SMC and SMC implemented in a SBBC prototype are also presented.

Keywords: sliding surface; sliding mode control; semi-bridgeless boost converter; adaptive hysteresis band; power factor correction; non-linear control

1. Introduction

Many electrical devices such as motors, computers and household appliances use passive rectifiers for supplying energy to DC loads. The rectifying action usually injects lower order harmonics which increase the Total Harmonic Distortion (THD_i) of the current, reduces the Power Factor (PF) and worsens the energy quality of electrical grids [1–3]. Active rectifiers have become an attractive alternative for overcoming these problems [4,5]. They are current-controlled rectifiers used to control the current in the AC side and provide a regulated DC voltage to load, their controllers are usually designed for keeping PF and THD_i within admissible ranges (PF > 0.9 and THD_i < 5%) according to IEEE Std. 519 and IEC/EN 61000-3-2 [6,7].

Several active rectifiers with PFC-based on boost converter have been proposed to replace passive rectifiers [8–10]. Among these, SBBC is a promising topology since it stands

Appl. Sci. **2021**, *11*, 1873. https://doi.org/10.3390/app11041873   https://www.mdpi.com/journal/applsci

out by reducing the number of diodes in the current path from source to load, decreasing conduction power losses and improving overall efficiency. In addition, SBBC topology has two clamped diodes that connect the source to the circuit ground, decreasing common mode noise and electromagnetic interference [8,11,12]. Due to the aforementioned reasons, the SBBC topology was selected as the topology under study of this paper.

SBBC controllers are usually based on classic cascade PI or linear controllers which feature an external voltage loop and an inner current loop that use Pulse Width Modulation (PWM) for generating the required control signals and activating power switches. However, this kind of controllers present start-up overvoltages causing instability and affecting sensitive loads [10,13–17]. Particularly, Kim et al. [17] made a comparative analysis for PFC of several high efficiency AC/DC boost topologies. They showed that linear controllers allow obtaining high PFs and low THD_is; however, there are some issues to highlight: (1) DC voltage response has over peaks; (2) AC current stabilization time has a delay of several cycles in face of disturbances and load changes; and (3) the AC current waveform near zero crossing presents distortions [9,11,14]. In general terms, linear PI controllers are designed around an operating point which reduces the dynamic response in presence of large disturbances; furthermore, the dynamic performance of the controller is degraded when the operation region of the converter moves away from the equilibrium point used in the design [8,11,13]. In addition, large disturbances and extensive changes in the operation point supremely affect sensitive loads [18]. To improve the response of power systems, the use of non-linear controllers with fast response under disturbances and high working range is desirable.

SMC is a non-linear control strategy that represents a good alternative for controlling variable structure systems, as SBBCs. SMC improves the robustness against large and fast disturbances and reduces the sensitivity of the system to the variation of parameters. Furthermore, SMC deals with uncertainty in modeling parameters. Moreover, it can directly provide the switching signals of power switches by means of hysteresis modulation. Consequently, the dynamic response in closed loop is the fastest possible [19–24].

The most relevant and recently published papers related to SBBC controllers, along with their main contributions and some drawbacks, are summarized hereafter.

In [12,25], the authors proposed a hybrid PI-SMC for a SBBC that allows reducing the injection of DC current into the power network. They implemented hysteresis modulation; nonetheless, the hysteresis band amplitude is constant, obtaining a variable switching frequency which increases THD_i. Their sliding surface function consists of reducing the error between voltage and current waveforms, guaranteeing AC current and voltage in phase. SMC ensures current control for PFC when load is increasing up to 140%; however, the DC voltage presents variations of up to 13% around the equilibrium point and a delay of several grid cycles for voltage stabilization is observed. Basically, DC voltage was not considered in the sliding surface design; hence, the controller response is slow for DC bus regulation.

In [26], the authors presented an analysis between integral and double integral SMC for AC current error. Their current sliding surface with an integral component presents low steady state error and high PF; nonetheless, the AC current has high THD_i. The sliding surface with double integral reduces steady state error and THD_i; however, a DC current component appears. For this sliding surface, DC voltage dynamic was not considered.

Sudalaimani et al. [27], Shieh and Chen [28] implemented a hybrid PI-SMC for current error reduction. In this case, the DC voltage is controlled by the external PI loop and it does not present over peaks when the load or source are disturbed or when the reference voltage is changed. Nevertheless, a slow response with delay of seconds for both AC current and DC voltage is observed.

Kessal and Rahmani [15] implemented an AC/DC converter with hybrid PI-SMC for PFC. In this case, a genetic algorithm (GA) for obtaining the sliding coefficients was developed. AC current and DC voltage errors were considered in the sliding surface design; however, the DC voltage response presents oscillations and high overvoltages in the presence of load disturbances since the SMC response was limited by the linear controller. In addition, THD_i increases due to the significant distortion produced in the AC current near zero crossing.

Mallik et al. [23] implemented an AC/DC converter based on totem-pole topology with a hybrid PI-SMC controller. AC current and DC voltage errors were considered in the sliding surface design. An integral surface of AC current and DC voltage errors were also included, reducing the steady state errors. In this case, fast responses were reached when AC current changes were incorporated; nevertheless, an external limiter for protection of power switches was required. In addition, the DC voltage presented over peaks due to the linear component of the controller.

Mohanty and Panda [18] presented an AC/DC converter with a hybrid PI-SMC controller and a PWM modulation for fixing switching frequency. The AC current and DC voltage with their corresponding integrative errors were considered. The DC voltage presented oscillations with over peaks of up to 11%, which were mitigated in several cycles of the source when the load changed, so that linear controller response was slower. Table 1 summarizes the key point of the literature review.

Table 1. Key points of the literature review.

Ref.	Contributions	Drawbacks
[12,25]	Hybrid PI-SMC with constant hysteresis band.	1. Variable switching frequency. 2. DC voltage not considered in the SS.
[15]	GA hybrid PI-SMC	1. DC voltage oscillations. 2. High DC overvoltages. 3. High THD_i near zero crossings.
[18]	Hybrid PI-SMC controller (fixed frequency).	1. DC voltage oscillations. 2. Slow linear control.
[23]	Hybrid PI-SMC controller (Totem-pole topology).	1. Protection external limiter required. 2. DC voltage over peaks.
[26]	Integral and double integral SMC.	1. Integral: high THD_i 2. Doble integral: DC current components. 3. DC voltage not considered in the SS. 4. DC voltage control was not implemented.
[27,28]	Hybrid PI-SMC for current error reduction.	1. AC-current and DC-voltage slow responses.

According to the reviewed literature, the knowledge gap consists of developing a SMC controller for SBBC that simultaneously performs Power Factor Correction (PFC) and DC bus regulation without using cascade PI or PI-SMC controllers to achieve the best dynamic behavior that removes DC overvoltages and responds fast to DC voltage changes or DC load current perturbations. For this, we propose the use of SMC with a new sliding surface with three terms that improves the following aspects: (1) A normalized DC voltage error term removes DC voltage instability in start-up and large disturbances, since the controller gives priority to the current controller component increasing robustness [29]. (2) An AC current error term implements a PFC scheme and guarantees fast current stabilization during disturbances. (3) An integral sliding term for the AC current error increases the overall system stability and reduces the stable-state error in presence of load or source disturbances. It is worth mentioning that this paper presents a rigorous deduction of the sliding mode conditions giving specific details and explaining the phenomenon of current waveform distortion that is presented in zero crossings.

Another contribution of the paper is the implementation of an Adaptive Hysteresis Band (AHB) to fix the switching frequency according to the system dynamic, which improves current waveform in zero crossing and reduces THD_i. The proposed AHB does not need PWM signals; hence, the SMC response is fast without compromising robustness.

This paper is organized as follows. Section 2 corresponds to the SMC design which includes transversality, existence and equivalent control conditions. Section 3 describes the controllers implemented in this paper: PI control, PI-SMC and SMC. Additionally, the AHB is described for PI-SMC and SMC. Section 4 presents the simulation results, including a comparative analysis of the control schemes and a detailed analysis of the proposed sliding surface illustrating its effectiveness. Section 5 presents the experimental results that demonstrate the effectiveness of the proposed SMC. Finally, conclusions are presented in Section 6.

2. Sliding Mode Control Design

In this section, the mathematical procedure for the SMC design is detailed through a five-step procedure: The first step consists of obtaining the SBBC mathematical model. The second step corresponds to the description of the proposed sliding surface for DC voltage regulation and PFC. The third step consists of the validation of the transversality condition. The fourth step is the validation of existence conditions. The fifth step is the validation of the equivalent control condition to evaluate SBBC under ideal control conditions. According to the authors of [30,31], transversality, existence and equivalent control conditions are sufficient for guaranteeing stability and convergence of the SMC.

The control objectives are as follows: (1) control the waveform of AC current, keeping it sinusoidal and in phase with AC voltage for PFC; (2) control the amplitude of the AC current ripple near zero crossing in order to reduce THD_i; and (3) regulate the DC voltage according to load requirements.

2.1. First Step: SBBC Mathematical Modeling

Figure 1 corresponds to the SBBC topology with clamped diodes. This topology has two inductors L_1 and L_2 (it is assumed that $L_1 = L_2 = L$); four diodes D_1, D_2, D_3 and D_4; two power switches Q_1 and Q_2; and a capacitor C. The AC source is denoted by $(v_s = V_s sin(wt))$, while the DC load is the resistor R. L_1, Q_1 and D_1 operate in the positive semi-cycle, while L_2, Q_2 and D_2 operate in the negative semi-cycle. SBBC can be modeled as a boost converter for each semi-cycle and switches actuate with the control signal $(u)(u = 1$ for closed switches and $u = 0$ for open switches). The mathematical model (Equations (1) and (2)) is obtained using Kirchhoff laws for both switching states. Equation (1) represents output voltage (v_o) dynamic (DC voltage) and Equation (2) describes the dynamic behavior of the AC current (i_s). For a detailed description concerning the operation principle, deduction of equations and non-minimum phase issues, the work of Mejía-Ruiz et al. [32] can be consulted.

$$\frac{dv_o}{dt} = -\frac{v_o}{RC} + (1-u)\frac{i_s}{C} \tag{1}$$

$$\frac{di_s}{dt} = \frac{v_s}{L} - (1-u)\frac{v_o}{L} \tag{2}$$

2.2. Second Step: Sliding Surface Proposal

The sliding surface (S) proposed in this paper is composed of the terms S_1, S_2 and S_3 as indicated in (3). Their respective sliding coefficients are α_1, α_2 and α_3. S_1 corresponds to the normalized DC voltage error where V_{ref} is the DC voltage established as a requirement in the design process [29]. S_2 corresponds to the AC current error, where the reference current (i_{ref}) is a rectified sinusoidal signal being I_{ref} its amplitude, as indicated by (4). I_{ref} is obtained through the power balance between SBBC input and output $(P = v_s i_{ref} = V_{ref} i_o)$. S_3 corresponds to the integral of the AC current error used to increase system stability. This integral reduces the amplitude of oscillations in start-up for input current or in face of large disturbances.

$$S = -\alpha_1 \underbrace{\left(\frac{v_o}{V_{ref}} - 1 \right)}_{S_1} - \alpha_2 \underbrace{(i_s - i_{ref})}_{S_2} - \alpha_3 \underbrace{\int (i_s - i_{ref})dt}_{S_3} \tag{3}$$

$$i_{ref} = \underbrace{\left(\frac{2V_{ref}v_o}{RV_s} \right)}_{I_{ref}} |sin(wt)| \tag{4}$$

Equations (1)–(3) are conveniently expressed using $x_1 = (v_o/V_{ref} - 1)$ and $x_2 = i_s - i_{ref}$ as follows:

$$\frac{dx_1}{dt} = \frac{(1-u)(x_2 + i_{ref})}{CV_{ref}} - \frac{(x_1 + 1)}{RC} \tag{5}$$

$$\frac{dx_2}{dt} = \frac{v_s}{L} - \frac{V_{ref}(1-u)(x_1 + 1)}{L} - \frac{di_{ref}}{dt} \tag{6}$$

$$S = -\alpha_1 x_1 - \alpha_2 x_2 - \alpha_3 \int x_2 dt \tag{7}$$

Equation (8) represents the reference current time derivative (di_{ref}/dt) where coefficients γ_1 and γ_2 are defined as auxiliary variables. Equations (6) and (8) can be combined for deducing the dynamics of current error in (9).

$$\frac{di_{ref}}{dt} = \underbrace{\left(\frac{2V_{ref}^2}{RV_s} |sin(wt)| \right)}_{\gamma_1} \frac{dx_1}{dt} - \underbrace{\left(\frac{2wV_{ref}^2}{RV_s} cos(wt) sign[sin(wt)] \right)}_{\gamma_2} (x_1 + 1) \tag{8}$$

$$\frac{dx_2}{dt} = \frac{v_s}{L} + \left[\frac{\gamma_1}{RC} - \gamma_2 - \frac{(1-u)V_{ref}}{L} \right](x_1 + 1) - \frac{\gamma_1(1-u)}{CV_{ref}}(x_2 + i_{ref}) \tag{9}$$

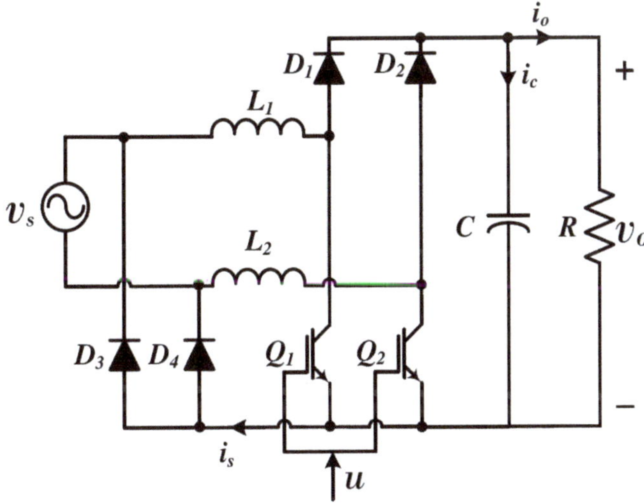

Figure 1. Topology SBBC with clamped diodes.

2.3. Third Step: Validation of Transversality Condition

Transversality condition allows evaluating if the SBBC can be controlled by using the control signal u according to (10). Equation (11) represents $\dot{S}$ that was obtained by deriving (7). The solution for the transversality condition is expressed by (12); this equation gives the first condition for sliding coefficients (α_1/α_2) (see (13)).

$$\frac{d}{du}\left(\frac{dS}{dt}\right) \neq 0 \tag{10}$$

$$\frac{dS}{dt} = \dot{S} = -\alpha_1\frac{dx_1}{dt} - \alpha_2\frac{dx_2}{dt} - \alpha_3 x_2 \tag{11}$$

$$\frac{d}{du}\left(\frac{dS}{dt}\right) = \frac{\alpha_1}{CV_{ref}}(x_2 + i_{ref}) - \alpha_2\left[\frac{V_{ref}}{L}(x_1 + 1) + \frac{\gamma_1}{CV_{ref}}(x_2 + i_{ref})\right] \neq 0 \tag{12}$$

$$\frac{\alpha_1}{\alpha_2} \neq \frac{CV_{ref}^2}{L}\frac{(x_1 + 1)}{(x_2 + i_{ref})} + \gamma_1 \tag{13}$$

Figure 2 is the representation of the right side of (13) (blue curve). Note that the maximum values tend to infinite when $wt = 0$ and $wt = \pi$, while the minimum value is presented with $wt = \pi/2$. Fulfilment of the transversality condition in the complete range of operation is given when the α_1/α_2 ratio is less than the minimum value of the curve (see (14)). According to Figure 2, (14) better expresses the condition of (13).

$$\frac{\alpha_1}{\alpha_2} < \frac{CRV_s}{2L} + \frac{2V_{ref}^2}{RV_s} \tag{14}$$

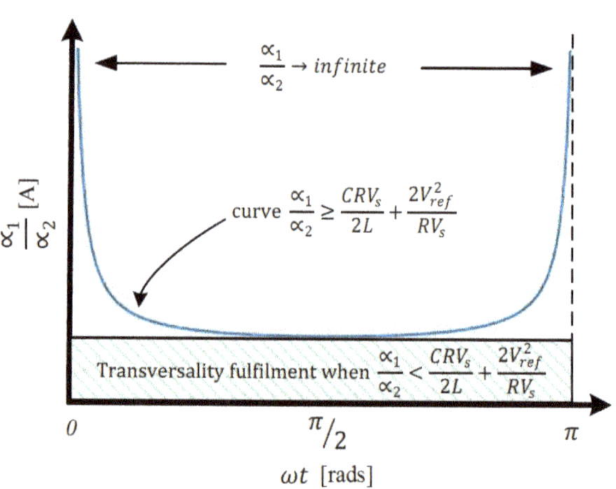

Figure 2. Graphical representation of transversality condition.

2.4. Fourth Step: Validation of Existence Condition

This condition ensures the existence of a sliding mode around $S = 0$, such that SBBC must remain within the sliding surface when $S \to 0$ (see (15)), guaranteeing not only $S = 0$ but also $x_1 = x_2 = 0$. Under these conditions, the system remains controlled around the equilibrium point.

$$\lim_{S=0^+}\frac{dS}{dt}\Big|_{u=1} < 0 \quad \text{and} \quad \lim_{S=0^-}\frac{dS}{dt}\Big|_{u=0} > 0 \tag{15}$$

Equation (16) presents the first existence condition when the system operates over the sliding surface and the control signal ($u = 1$) is applied for following the dynamic trajectories towards $S = 0$.

$$\lim_{S=0^+}\frac{dS}{dt}\Big|_{u=1} = \left[\frac{\alpha_1}{\alpha_2}\frac{1}{RC} + \gamma_2 - \frac{\gamma_1}{RC}\right](x_1 + 1) - \frac{\alpha_3}{\alpha_2}x_2 - \frac{v_s}{L} < 0 \tag{16}$$

Figure 3a is the representation of (16), while Figure 3b is the zoom in at the extremes ($wt = 0$ and $wt = \pi$) which correspond to zero crossings. Note that dS/dt is lower than zero for almost the entire operating range, while dS/dt is greater than zero around zero crossings. Thus, for the entire operating range, regions near zero crossings correspond to unstable points and the SBBC cannot be controlled. In a strict sense of sliding mode theory, this condition is not fulfilled and sliding mode control should not be implemented. However, in practical terms, the instability of the system is produced in a short period of time, and basically this is the reason AC current deforms at zero crossing. Zero crossing deformation is not exclusive to the theory of sliding modes; this phenomenon also occurs when other control techniques are implemented [9,11,14]. In general terms, sliding mode theory allows explaining the phenomenon of zero crossing deformation. To mitigate the deformation at zero crossings, an adaptive hysteresis band based on current geometry is proposed, as described in Section 3.

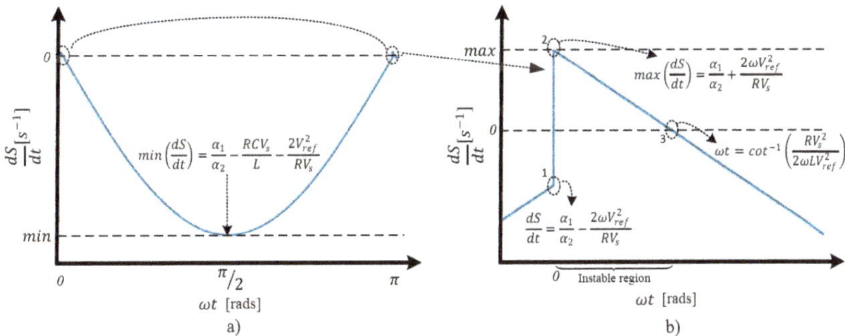

Figure 3. dS/dt for existence condition when $u = 1$. (**a**) Representation in the range of $wt = [0, \pi]$; (**b**) Zoom in at $wt = 0$ and $wt = \pi$.

Equation (17) provides the second condition for α_1/α_2 which was obtained by evaluating (16) in system boundaries and in steady state. This condition is more restrictive than the transversality one.

$$\frac{\alpha_1}{\alpha_2} < \frac{2wCV_{ref}^2}{V_s} \tag{17}$$

Equation (18) presents the second existence condition when the system operates under the sliding surface, and the open control signal ($u = 0$) is applied to follow the dynamic trajectories towards $S = 0$. Figure 4 shows that dS/dt is greater than zero for the entire operating range. This condition is obtained considering (17).

$$\lim_{S=0^-} \frac{dS}{dt}\Big|_{u=0} = \left(\frac{\alpha_1}{\alpha_2} - \gamma_1\right)\left[\frac{(x_1+1)}{RC} - \frac{(x_2 + i_{ref})}{CV_{ref}}\right] +$$
$$\left(\frac{V_{ref}}{L} + \gamma_2\right)(x_1 + 1) - \frac{V_s}{L} - \frac{\alpha_3}{\alpha_2}x_2 > 0 \tag{18}$$

Another condition that is usually verified in SMC is the hitting condition. This one is related to the existence condition and can be satisfied if the switching function is appropriately chosen [33,34]. Equation (19) corresponds to the switching function proposed in this paper. Satisfying this condition assures that, regardless of the initial condition, dynamic trajectories of the system are always towards the sliding surface. The suitable selection of the logic states of the switching function and its application in the SM control law allow changing the dynamic behavior of the system. As a consequence, for the design of the hitting condition, it is sufficient to consider state variables x_1 and x_2, during the reaching phase. If the measured magnitude of i_s is less than i_{ref} and v_o is less than v_{ref}, then S is positive. In this condition, the necessary switching action required to force the system trajectories towards the sliding surface $S = 0$ is $u = 1$, causing $\dot{S} < 0$. When the power switch is ON, the inductor L stores the energy supplied by the source, increasing the magnitude of i_s until $S = 0$. Otherwise, when the power switch is OFF, the induced voltage across L is added to v_s, supplying power to the capacitor and reducing the magnitude of i_s with respect to i_{ref}. Accordingly, the hitting condition is closely related to the way in which the switching states and hysteresis band are selected, as exhibited in (19).

$$u(S) = \frac{1}{2}(1 + sign(S)) = \left\{ \begin{array}{llll} 1 & when & S > 0 & and & \dot{S} < 0 \\ 0 & when & S < 0 & and & \dot{S} > 0 \end{array} \right\} \tag{19}$$

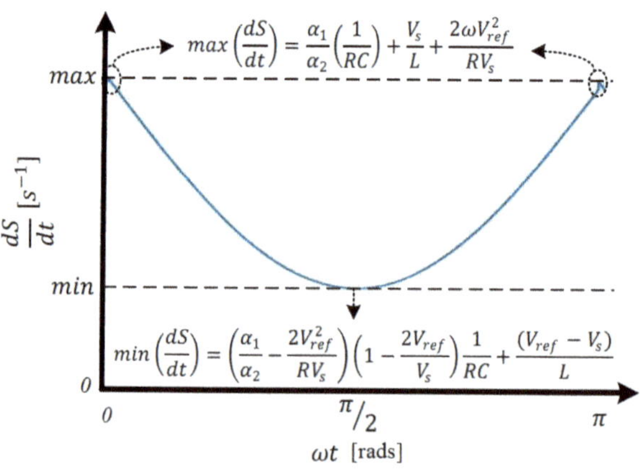

Figure 4. dS/dt for existence condition when $u = 0$.

2.5. Fifth Step: Validation of Equivalent Control

The equivalent control evaluates the system dynamics under ideal operation conditions ($x_1 = x_2 = 0$), assuming an infinite switching frequency and disregarding the time variation of the sliding surface ($\dot{S} = 0$). Equation (20) presents the equivalent control condition (u_{eq}) obtained from (5), (9) and (11). The condition for ensuring the equivalent control ($0 < u_{eq} < 1$) is presented in (21) and corresponds to the third condition for (α_1/α_2) evaluated at the system boundaries. In this case, this condition corresponds to the existence constraint found in (17), which also ensures that $0 < u'_{eq} < 1$ for the entire range of operation, except in zero crossings, as indicated in Figure 5.

$$u_{eq} = 1 - \underbrace{\frac{\left[\frac{\alpha_1}{\alpha_2} + RC\gamma_2 - \gamma_1\right]\frac{(x_1+1)}{RC} - \frac{\alpha_3}{\alpha_2}x_2 - \frac{v_s}{L}}{\left(\frac{\alpha_1}{\alpha_2} - \gamma_1\right)\frac{(x_2+i_{ref})}{CV_{ref}} - \frac{V_{ref}}{L}(x_1 + 1)}}_{u'_{eq}} \tag{20}$$

$$\frac{\alpha_1}{\alpha_2} < \frac{2wCV_{ref}^2}{V_s} \tag{21}$$

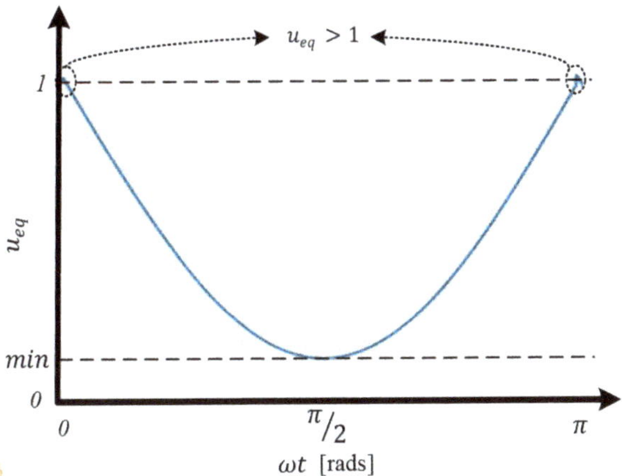

Figure 5. u_{eq} for half cycle of electrical grid.

3. Control Schemes

Figure 6 presents the SBBC control system. The voltage waveform for obtaining the reference waveform of AC current is taken from the electrical grid by means of Phase-Locked Loop (PLL). The control system requires the measurement of v_s, i_s, v_o and i_o signals. v_o and i_o are filtered to remove high frequency noise. Finally, the controller calculates the control actions and generates the u signal to trigger power switches Q_1 and Q_2.

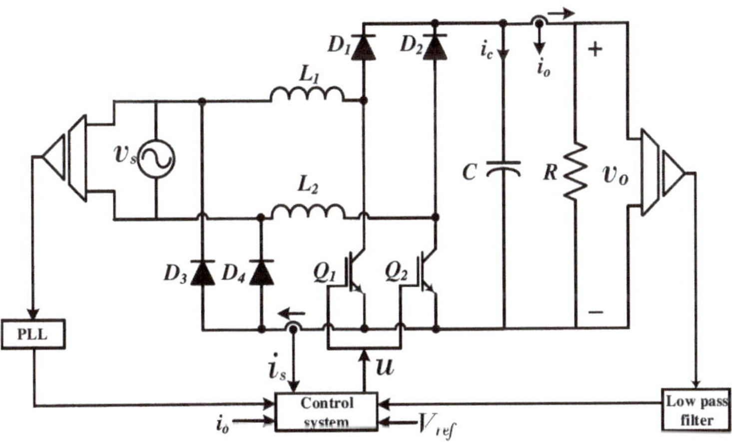

Figure 6. Control system for SBBC.

Three control strategies are evaluated: PI, hybrid PI-SMC and the proposed SMC (see Figure 7). A classical cascade PI controller is shown in Figure 7a. The external voltage loop gives the AC current amplitude reference to internal current loop and current controller actuates by means of PWM generator. The control Scheme of the hybrid PI-SMC is depicted in Figure 7b; in this case, there is an external PI control loop for DC voltage regulation that gives the reference AC current amplitude for the internal SMC; the voltage error is considered by the SMC and an AHB is used instead of a PWM generator. The proposed SBBC control with SMC is presented in Figure 7c. This system allows controlling both DC voltage and AC current; the reference current is obtained from the power balance and the adaptive hysteresis band gives the control signal. The proposed sliding surface is used in both hybrid PI-SMC and SMC controllers in order to analyze their advantages and limitations.

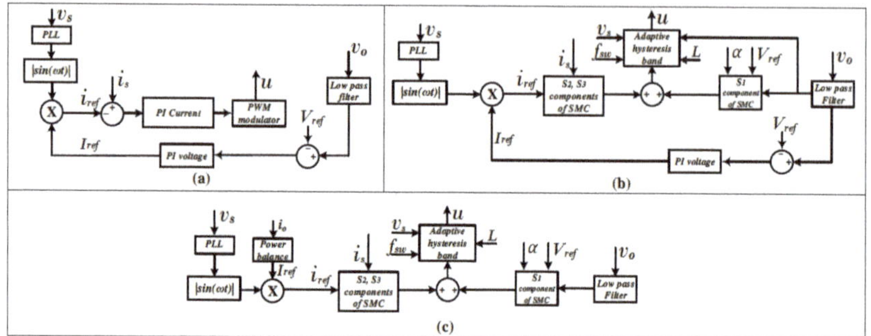

Figure 7. Block diagram of the control system: (**a**) classical cascade PI controller; (**b**) hybrid PI-SMC; and (**c**) proposed SMC.

The AHB is designed based on current ripple geometry according to Mejía-Ruiz et al. [35]. It has two functions: (1) smooth the current zero crossing in order to decrease THD_i; and (2) modify the switching time according to SBBC operation point in order to fix the switching frequency. Equation (22) presents the AHB as function of v_o, v_s, L and f_{sw}, while its implementation is depicted in Figure 8. The AHB calculated for each point is compared with the sliding surface. Finally, the control signal is given by a RS flip-flop. The AHB does not require PWM signals for actuating over power switches; thus, sliding mode control does not have any delay in its AC current response.

$$AHB = \frac{v_s(v_o - v_s)}{2L f_{sw} v_o} \tag{22}$$

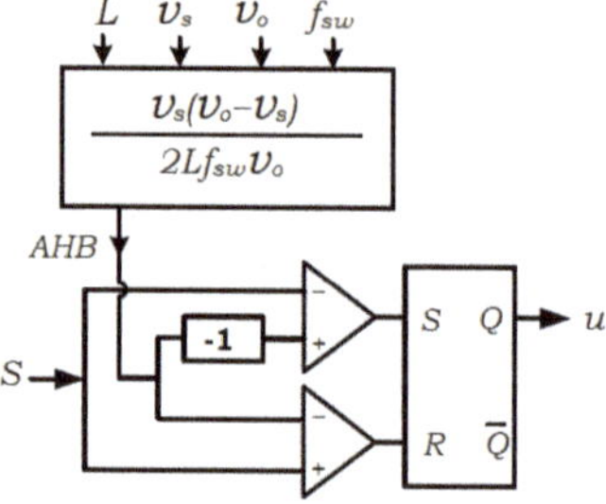

Figure 8. Adaptative hysteresis band modulator.

4. Simulation Results

The proposed sliding surface and its control were performed in PSIM computer simulation software. The comparative analysis for PI, hybrid SMC-PI and the proposed SMC control strategies is presented in Section 4.1, while the surface dynamic behavior at start-up is discussed in Section 4.2. The values of the system parameters are given in Table 2. The PI control parameters used in simulations are $K_p = 0.2$ and $K_i = 0.2$ for the voltage loop and $K_p = 104$ and $K_i = 0$ for the current loop. The stability analysis and control tuning were performed in sisotool (MATLAB R2020a) by means of root locus analysis together with the "Robust Response Time" tuning method. The tuning steps and criteria such as voltage overshot lower than 10% and delay between internal and external loops can be consulted in [32]. In addition, the coefficient for SMC is $\alpha = 150$, which is in the stable admissible range.

Table 2. Simulation parameters.

Parameter	Value
Grid voltage $(v_s = v_{in})$	120 Vrms
Grid frequency (f)	60 Hz
DC bus capacitor (C)	2.2 mF
DC bus voltage $(V_o = V_{ref})$	400 V
Inductors $(L_1 = L_2 = L)$	2.2 mH
Switching frequency (f_{sw})	40 kHz
Rated power (P_o)	500 W
Rated Load (R)	320 Ω

4.1. Comparative Analysis for PI, PI-SMC and SMC

The effect of load increase on SBBC performance with PI, PI-SMC and the proposed SMC controllers is analyzed. In this simulation, the DC power of the load was increased from 375 to 500 W for PI and PI-SMC controllers. In addition, to better show the performance of the SMC controller, the load was increased from 250 to 500 W. Figure 9 compares the AC current stabilization time for the three controllers. The AC current with PI controller exhibits a stabilization time of 0.4 s. This control system presents a delay between measurement and switching signal significantly degrading the waveform near zero crossing and presenting a THD_i of 17.23% (Figure 9a). For PI-SMC, the hysteresis modulation strategy improves the waveform near zero crossing of current, reducing THD_i to 3.85% (Figure 9b); nevertheless, the current stabilization (amplitude) is reached after 24 cycles, the same time as the PI controller. Finally, The proposed SMC allows regulating the AC current, rapidly responding to operation changes in only 0.04 ms. In this case, it can be observed that the current response is the fastest (Figure 9c). It can be concluded that SMC and the use of an ABH allow decreasing the THD_i to 3.7% and SMC presents the fastest response under perturbations.

Figure 10 shows the DC voltage response. The power converter with PI control (Figure 10a) takes 395.7 ms to regulate the DC voltage and presents an overvoltage of 12.75%. In comparison, PI-SMC allows reducing the overvoltage to 3.75%, maintaining the same stabilization time of the PI control (Figure 10b). This result evidences the delay caused by the PI control when it is used as external loop, limiting the transient response of the closed loop system. On the other hand, the DC voltage stabilizes in 30 ms, when the SMC strategy is used and the response exhibits a low overvoltage of 0.1% (Figure 10c). This evidences that SMC responds rapidly, protecting sensitive DC loads against overvoltages.

4.2. Sliding Mode Control Behavior

This section presents the stability of SMC, forcing the system to work in adverse conditions consisting on the start-up without a pre-charge of the DC bus or under large perturbations.

The response of the DC voltage in SBBC start-up is presented in Figure 11. Figure 11a shows the DC voltage without DC-bus pre-charge. Note that v_o does not present instability or over peaks and it increases until reaching the stabilization point (400 V) in 1.52 s. The stabilization time can be modified by means of the sliding coefficient depending on the desired priority order, increasing or decreasing the response time for DC voltage or AC current. Figure 11b presents the behavior of the normalized DC voltage error (S_1); initially, voltage error is bounded by its minimum value (-1). Therefore, S_1 at start-up according to the sliding coefficient value. Then, S_1 rapidly reaches the convergence point (stabilization point) and keeps sliding around $S_1 = 0$. Normalization of output voltage error allows reducing the output voltage impact during large disturbances, giving priority to current control and avoiding non-minimum phase behavior [29].

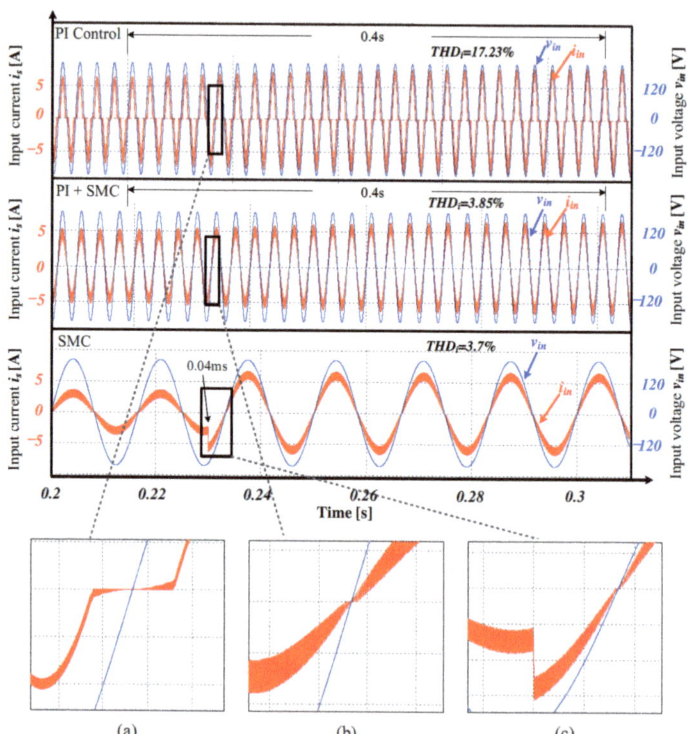

Figure 9. AC current performance of: (**a**) PI control; (**b**) PI-SMC; and (**c**) the proposed SMC.

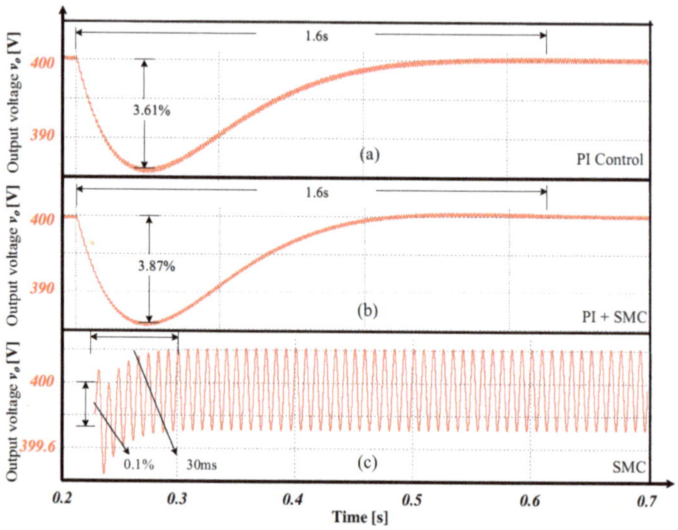

Figure 10. DC voltage transient response when load increases from 250 W to 500 W, comparing the performance of (**a**) PI control; (**b**) PI-SMC; and (**c**) SMC.

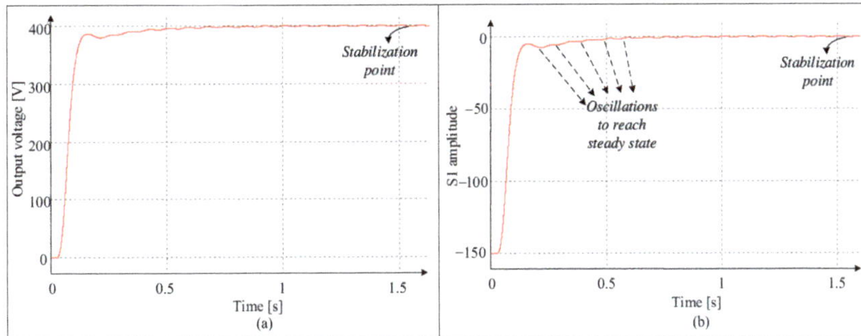

Figure 11. DC voltage behavior in start-up: (**a**) DC voltage; and (**b**) DC voltage error S_1.

The behavior of AC current in SBBC in start-up is presented in Figure 12. In Figure 12a, a transient with an over peak near 15 times the desired value is observed; nevertheless, the over peak is only presented for the first half cycle of the wave and the AC current rapidly stabilizes in a sinusoidal wave in the second grid cycle while zero crossing deformation is gradually reduced in four cycles. Figure 12b presents the behavior of AC current error (S_2). Initially, S_2 also has a couple of large transient oscillations, and then the magnitude of oscillations is significantly reduced, being only evident for the first zero crossings. S_2 reaches the convergence zone oscillating near the equilibrium point in the adaptive hysteresis band. The main function of S_2 consists on properly following the AC current reference according to load requirements which permits avoiding the non-minimum phase behavior; in addition, another function of S_2 consists on performing PFC.

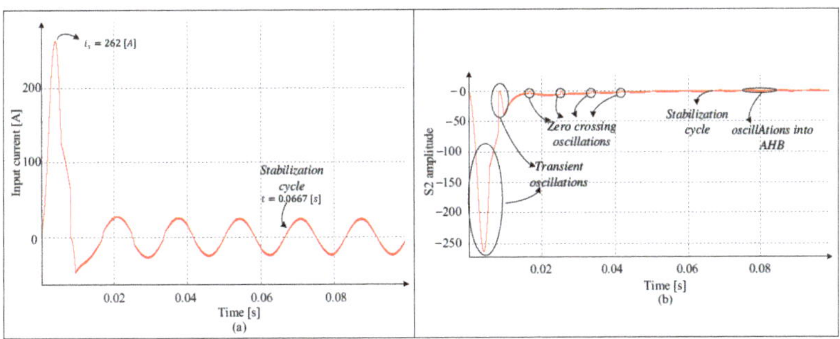

Figure 12. AC current behavior in start-up: (**a**) AC current; and (**b**) AC current error S_2.

The behavior of S_3 in start-up is presented in Figure 13. S_3 is reset each electric grid cycle to set the cumulative error as zero. S_3 begins in zero and immediately decreases near -1.2 for the first peak; then, it increases again and reaches the convergence zone around $S_3 = 0$. S_3 provides robustness under disturbances; its effect is more evident during start-up reducing oscillations before reaching the steady state.

To illustrate the impact of S_3 during start-up, Figure 14 shows the input current without using S_3. Comparing Figures 12a and 14, it can be seen that current overshoot and stabilization time are reduced when S_3 is included in the sliding surface. The stabilization of DC voltage and AC current and the convergence to zero of their errors (including integral) in SBBC start-up demonstrate the stability of the proposed sliding surface outside the operation zone (hitting condition and Lyapunov criteria).

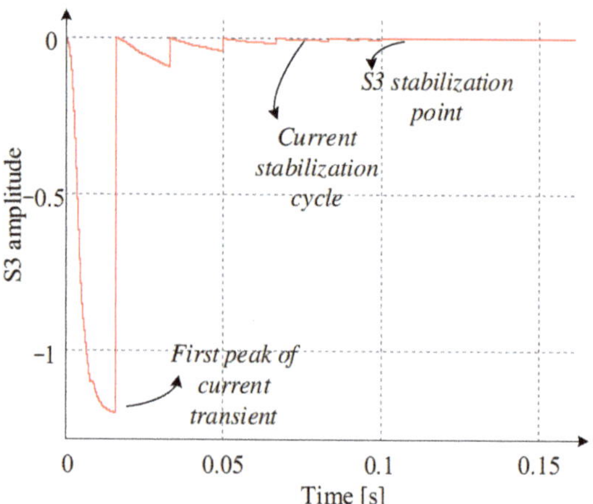

Figure 13. Integral of current error in start-up.

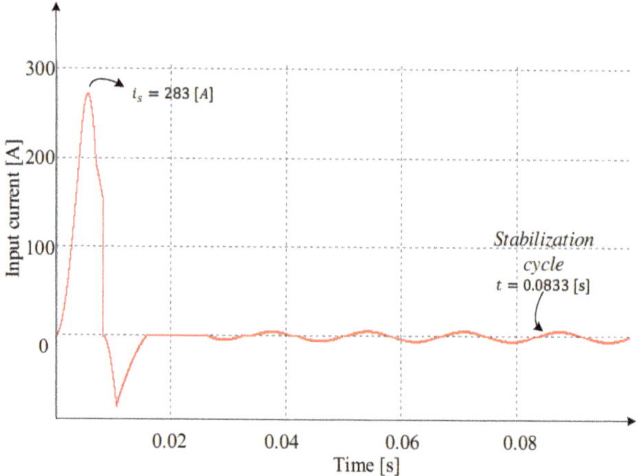

Figure 14. Input current behavior without integral of current error in start-up.

The sliding surface dynamic behavior when a disturbance appears is shown in Figure 15. The disturbance corresponds to an increase to twice the DC load current. Figure 15 presents the sliding surface limited by the AHB in half-cycle of the AC voltage. In this case, the system is forced to be above the upper AHB, and the SMC response immediately conducts the system within the AHB, which validates the hitting condition that is observed. Concerning the existence condition, the AHB limits the speed of the response, fixing the switching frequency at 40 kHz and forcing the system to be sliding around $S = 0$.

In addition, at the beginning and end of each half-cycle during zero crossings, the system is also out of the AHB; however, the SMC reacts and rapidly brings the system back into the AHB, which significantly reduces the THD_i produced in zero crossings.

Figure 16 depicts the frequency spectrum of AC current, using the proposed SMC. The energy is concentrated near the fundamental frequency (60 Hz) and the switching frequency (40 kHz). In this case, THD_i is 3.7%.

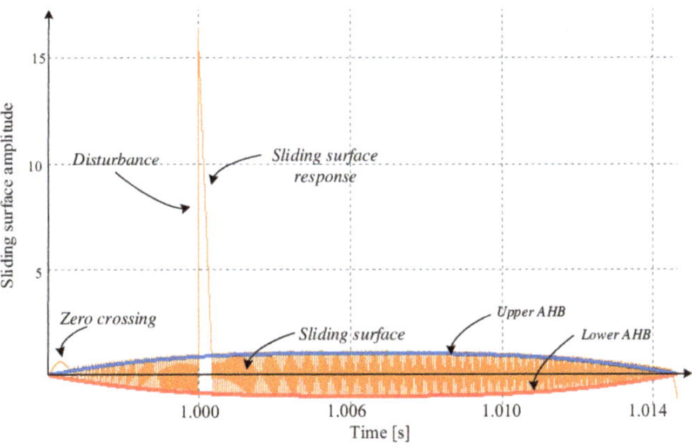

Figure 15. Sliding surface response for load increase.

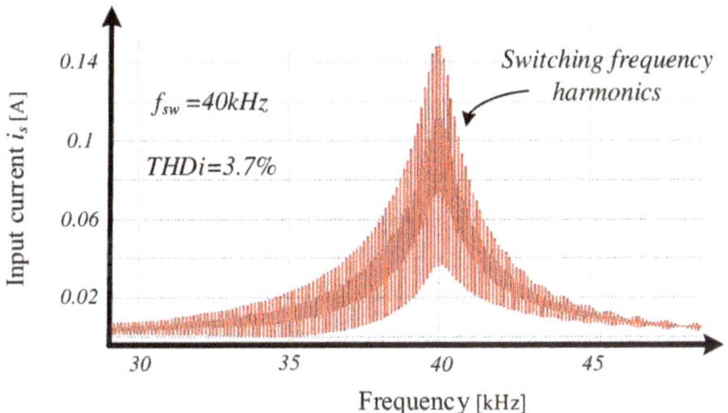

Figure 16. Frequency spectrum of the AC current using the adaptive hysteresis band.

5. Experimental Results

This section presents the experimental results when the hybrid PI-SMC (Section 5.1) and the proposed SMC (Section 5.2) controllers are implemented in a SBBC. Cascade PI control was not considered since it is presented in numerous previous works and the simulations show that this controller has the worst performance. Section 5.3 shows a zero crossing comparison between PI-SMC and SMC.

The SBBC implementation is presented in Figure 17. This prototype is mainly composed of: (1) the AC supply terminals and protection system; (2) Microcontroller Printed Circuit Board (PCB) used for signal processing and control (TMS 329F28335ZJZA); (3) measurement PCBs (ACS714 and ADUM5010) for measuring AC voltage, AC current and DC voltage; (4) EMI filter for improving electromagnetic compatibility; (5) inductors L_1 and L_2 for coupling AC and DC systems; (6) power switching PCB with power switches (IGBTs) and diodes (G4PC50UD); and (7) DC bus for regulating DC voltage. SBBC Requirements and characteristics are the same as those used in the simulation results section (Table 2). A Digital scope GW Instek GDS-2204A was used (200 MHz Bandwidth, 4 Input Channel, 2 GSa/s Real-time Sampling Rate, 2 Mp Record Length) for collecting the experimental results.

The project was completely developed in C and its coding developed in Code Composer Studio (CCS) software. The control algorithm runs in two main stages: (1) i_s, v_s, i_0 and v_0 are sampled with 12 bits of resolution, using the Analog-to-Digital Conversion (ADC) peripheral, the computation of the sliding surface according to Equation (3) and the generation of the trigger signal u. The execution time of these control routines is regulated based on an interruption of the CPU timer of the DSP F28335 every 1 μs. (2) The

PLL algorithm is executed, the hysteresis band computation according to Equation (22) is completed, the output voltage filtering is executed to reduce the 120 Hz ripple using a notch filter and i_{ref} is computed based on Equation (4). When the hybrid controller is run, the PI control is also computed in this routine.

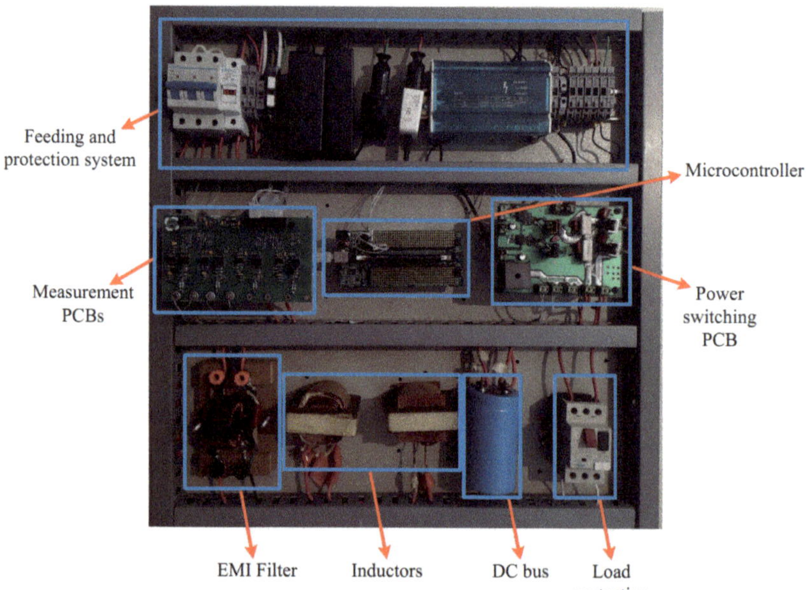

Figure 17. Semi-bridgeless boost converter for experimental tests.

5.1. Results for PI-SMC Controller

Figure 18 presents the results when the set point of the DC voltage is changed. Initially, it was set in 400 V and a change of 20 V was made, reducing DC voltage from 400 to 380 V, as shown in Figure 18a. Then, the DC voltage set point was increased from 380 to 400 V (Figure 18b). In both cases, the DC voltage presents a stabilization time of 1.3 s with overvoltage around the stabilization point. In addition, the current amplitude slightly changes. Nonetheless, the controller keeps the AC current and voltage in phase, ensuring a *PF* close to 1.

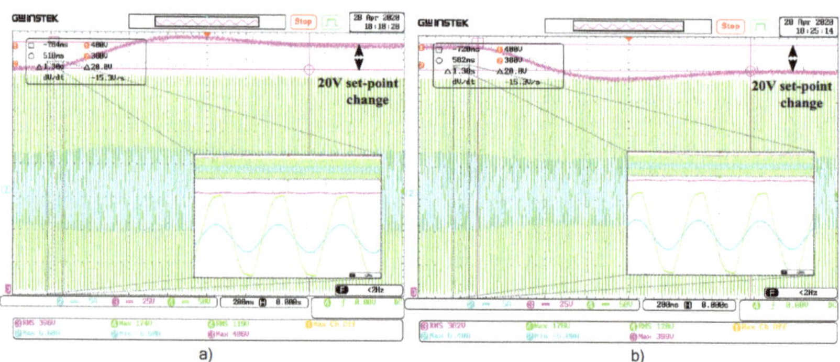

Figure 18. Experimental results with PI-SMC control for voltage set point changes: (**a**) DC voltage increases 20 V; and (**b**) DC voltage decreases 20 V.

Figure 19 shows the results when a load disturbance is caused. Figure 19a presents results when load decreases 25%. In this case, an overvoltage of 22 V and a DC voltage stabilization time of 2.9 s are observed. Figure 19b depicts the results when the load increases 25%; SBBC presents a maximum oscillation of 18 V and the DC bus stabilizes at 1.53 s. In both tests, the AC current presents a delay of several grid cycles before reaching the stabilization point.

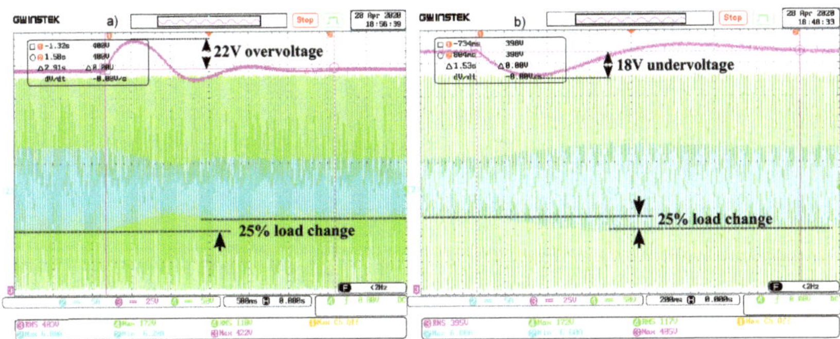

Figure 19. Experimental results with PI-SMC control for load changes: (**a**) load decreases 25%; and (**b**) load increases 25%.

5.2. Results for the Proposed SMC Controller

Similar to the previous section, Figure 20 presents the results when the set point of the DC voltage is changed. Initially, it was set in 400 V and a change of 20 V was made, reducing the DC voltage from 400 to 380 V, as shown in Figure 20a. Then, the set point was increased from 380 to 400 V (Figure 20b). In both cases, the DC voltage presents a stabilization time of 1.55. In this case, it does not have any overvoltage and the *PF* is ensured to be close to 1. In addition, the AC current amplitude is reached in only half a cycle after the set point change; in contrast, with the PI-SMC control, the AC current amplitude is reached after several cycles.

The violet waveform in Figure 20 corresponds to v_o. This result confirms that the SM controller effectively works and quickly reaches the new steady-state conditions in the presence of a significant change in operating conditions. This measure confirms the stability and robustness of the control technique proposed in this paper.

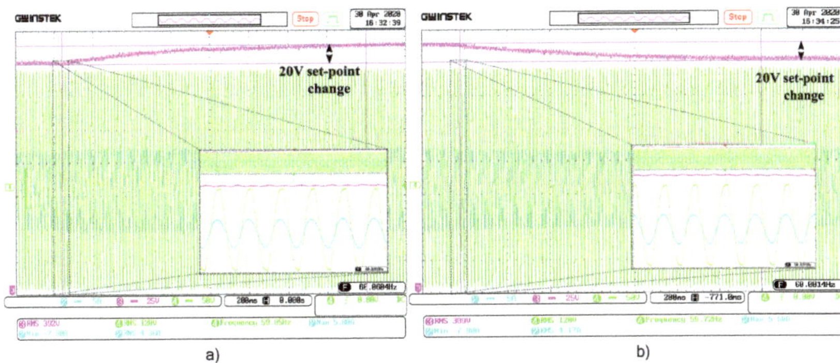

Figure 20. Experimental results with SMC for voltage set point changes: (**a**) DC voltage increases 20 V; and (**b**) DC voltage decreases 20 V.

Figure 21 depicts the results for load perturbations; particularly, Figure 21a presents the results when load decreases 50%, while Figure 21b corresponds to a load increment of

50%. In this case, the DC voltage does not present any overvoltage or oscillations around its set point. In addition, the SBBC instantaneously responds to load changes and the current amplitude is reached in the next half-cycle. The proposed SMC control presents the fastest response among the control strategies implemented, even when the load disturbance is twice as large as the load disturbance of PI-SMC control.

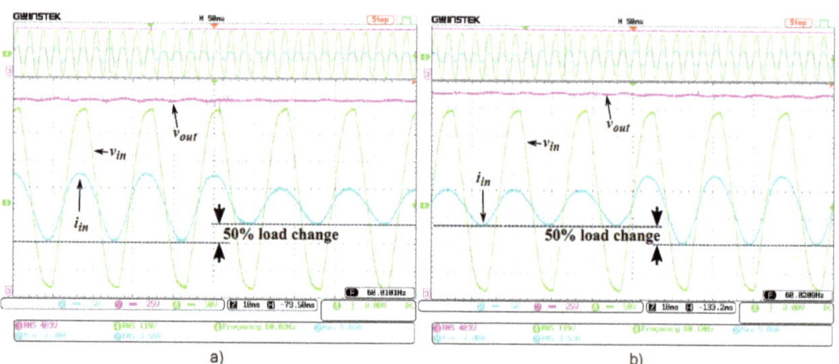

a)   b)

Figure 21. Experimental results with SMC for load changes: (**a**) load increases to 50%; and (**b**) load decreases to 50%.

5.3. Zero Crossing Comparison

A zoom for AC current results is shown in Figure 22 for the following two cases: (a) the PI-SMC with fixed hysteresis band; and (b) the SMC with the proposed AHB. In Figure 22a, ripple increments are observed just before and after the zero crossing obtaining a $THD_i = 4.83\%$. In Figure 22b, the zero crossing is softer than the one with fixed hysteresis band. This approach represents a better performance during zero crossing and a reduction of THD_i of 2.67%.

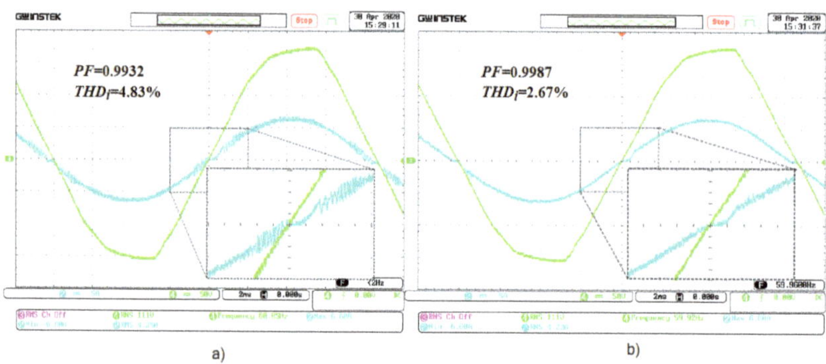

a)   b)

Figure 22. Zero crossing of AC current: (**a**) PI-SMC control; and (**b**) SMC control.

5.4. Experimental vs. Simulation Results

Table 3 presents some comparisons between simulation and experimental results when load increases: (1) for Hybrid SMC-PI, the load is increased 25%; and (2) for the proposed SMC, the load is increased 50%. The stabilization time and undervoltage for v_o, the cycles to reach steady state for i_s and its corresponding THD_i were compared. It can be observed that there are small differences for both cases.

Table 3. Experimental vs. simulation: test with load increase.

Criteria	Hybrid SMC-PI (Load 25%)		Proposed SMC (Load 50%)	
	Simulation	Experimental	Simulation	Experimental
v_o Stabilization time	1.6 s	1.53 s	0.03 s	0 s
v_o Undervoltage	15.5 v	18 v	0.4 v	0 v
i_s Cycles to reach steady state	24	27	less than 1	less than to 1
THD_i	4.25%	4.82%	3.7%	2.67%

6. Conclusions

This paper proposes a new sliding surface for controlling a SBBC which simultaneously incorporates PFC and DC bus regulation. The proposed sliding surface contains: (1) a normalized DC voltage error term; (2) an AC current error term; and (3) an integral of the AC current error term. The surface was validated using sliding mode conditions, simulations and experimental tests. The surface was implemented for PI-SMC and SMC, and an AHB was used for fixing the switching frequency and reducing THD_i.

Simulations were performed to compare the three controllers Cascade PI, PI-SMC and SMC in terms of their dynamic behavior. When applied a DC power change of 250 W (50%), it was found that SMC presented the best performance, since the DC voltage presents the lowest stabilization time (30 ms) and practically does not present overvoltage (0.1%) which guarantees the protection of sensitive loads. In addition, the AC current has the fastest time response (0.04 ms) and presents the best behavior in zero crossings which reduces THD_i (3.7%).

Several simulations were also carried out concerning the behavior of the proposed sliding surface using SMC in start-up and without a pre-charge of the DC bus, forcing the system to work in adverse conditions. In this situation, the SMC responded adequately limiting DC overvoltages and stabilizing the AC current within the first half cycle despite the large start-up over current. In these simulations, the stability of the SBBC was evidenced when the SMC was implemented.

Experiments were implemented for PI-SMC and SMC considering changes in the DC voltage set-point and in the DC load current. It was found that the SBBC responds better under changes and perturbations, even though the perturbations of the SMC were more severe than the ones of the PI-SMC. Finally, zero crossing comparison showed a better behavior for SMC working with AHB, which demonstrates the rapid response of SMC under instantaneous sign changes of the sliding surface.

Author Contributions: Conceptualization, J.R.O.-C., G.E.M.-R., N.M.-G. and J.B.C.-Q.; Data curation, J.R.O.-C.; Formal analysis, J.R.O.-C., G.E.M.-R., N.M.-G. and J.B.C.-Q.; Funding acquisition, N.M.-G. and J.M.L.-L.; Investigation, J.R.O.-C., G.E.M.-R. and N.M.-G.; Methodology, J.R.O.-C., G.E.M.-R., N.M.-G. and J.B.C.-Q.; Project administration, N.M.-G.; Resources, J.R.O.-C. and G.E.M.-R.; Software, J.R.O.-C. and G.E.M.-R.; Supervision, N.M.-G. and J.M.L.-L.; Validation, J.R.O.-C. and G.E.M.-R.; Visualization, J.R.O.-C., N.M.-G. and J.M.L.-L.; Writing—original draft, J.R.O.-C.; and Writing—review and editing, G.E.M.-R., N.M.-G., J.M.L.-L. and J.B.C.-Q. All authors have read and agreed to the published version of the manuscript.

Funding: This research was funded by the Colombia Scientific Program within the framework of the so-called Ecosistema Científico (Contract No. FP44842-218-2018).

Institutional Review Board Statement: Not applicable.

Informed Consent Statement: Not applicable.

Data Availability Statement: The data presented in this study are openly available in Github repository (https://github.com/joseroca1989/appendix-paper-SMC, accessed on 19 January 2021).

Acknowledgments: The authors thank to Corporacion Universitaria Minuto de Dios—UNIMINUTO for the support received in this research. The authors also acknowledge the support from the

Colombia Scientific Program within the framework of the call Ecosistema Científico (Contract No. FP44842-218-2018).

Conflicts of Interest: The authors declare that they have no conflict of interest.

References

1. Baek, J.; Kim, J.; Lee, J.; Park, M.; Moon, G. A New Standby Structure Integrated With Boost PFC Converter for Server Power Supply. *IEEE Trans. Power Electron.* **2019**, *34*, 5283–5293. [CrossRef]
2. Xu, H.; Chen, D.; Xue, F.; Li, X. Optimal Design Method of Interleaved Boost PFC for Improving Efficiency from Switching Frequency, Boost Inductor, and Output Voltage. *IEEE Trans. Power Electron.* **2019**, *34*, 6088–6107. [CrossRef]
3. Bist, V.; Singh, B. An Adjustable-Speed PFC Bridgeless Buck–Boost Converter-Fed BLDC Motor Drive. *IEEE Trans. Ind. Electron.* **2014**, *61*, 2665–2677. [CrossRef]
4. Cho, K.S.; Lee, B.K.; Kim, J.S. CRM PFC Converter with New Valley Detection Method for Improving Power System Quality. *Electronics* **2020**, *9*, 38. [CrossRef]
5. Zhang, R.; Ma, W.; Wang, L.; Hu, M.; Cao, L.; Zhou, H.; Zhang, Y. Line Frequency Instability of One-Cycle-Controlled Boost Power Factor Correction Converter. *Electronics* **2018**, *7*, 203. [CrossRef]
6. IEEE. *Recommended Practice and Requirements for Harmonic Control in Electric Power Systems (IEEE Std 519-2014 (Revision of IEEE Std 519-1992))*; IEEE: Piscataway, NJ, USA, 2014; pp. 1–29.
7. IEEE. *IEEE Recommended Practice—Adoption of IEC 61000-4-15:2010, Electromagnetic Compatibility (EMC)—Testing and Measurement Techniques—Flickermeter-Functional and Design Specifications (IEEE Std 1453-2011)*; IEEE: Piscataway, NJ, USA, 2011; pp. 1–58.
8. Huang, Q.; Huang, A.Q. Review of GaN totem-pole bridgeless PFC. *CPSS Trans. Power Electron. Appl.* **2017**, *2*, 187–196. [CrossRef]
9. Huber, L.; Jang, Y.; Jovanovic, M.M. Performance Evaluation of Bridgeless PFC Boost Rectifiers. *IEEE Trans. Power Electron.* **2008**, *23*, 1381–1390. [CrossRef]
10. Musavi, F.; Edington, M.; Eberle, W.; Dunford, W.G. Evaluation and Efficiency Comparison of Front End AC-DC Plug-in Hybrid Charger Topologies. *IEEE Trans. Smart Grid* **2012**, *3*, 413–421. [CrossRef]
11. Alam, M.; Eberle, W.; Gautam, D.S.; Botting, C. A Soft-Switching Bridgeless AC–DC Power Factor Correction Converter. *IEEE Trans. Power Electron.* **2017**, *32*, 7716–7726. [CrossRef]
12. Marcos-Pastor, A.; Vidal-Idiarte, E.; Cid-Pastor, A.; Martínez-Salamero, L. Loss-Free Resistor-Based Power Factor Correction Using a Semi-Bridgeless Boost Rectifier in Sliding-Mode Control. *IEEE Trans. Power Electron.* **2015**, *30*, 5842–5853. [CrossRef]
13. Tan, F.D.; Ramshaw, R.S. Instabilities of a boost converter system under large parameter variations. *IEEE Trans. Power Electron.* **1989**, *4*, 442–449. [CrossRef]
14. Fan, J.W.; Yeung, R.S.; Chung, H.S. Optimized Hybrid PWM Scheme for Mitigating Zero-Crossing Distortion in Totem-Pole Bridgeless PFC. *IEEE Trans. Power Electron.* **2019**, *34*, 928–942. [CrossRef]
15. Kessal, A.; Rahmani, L. Ga-Optimized Parameters of Sliding-Mode Controller Based on Both Output voltage and Input Current With an Application in the PFC of AC/DC Converters. *IEEE Trans. Power Electron.* **2014**, *29*, 3159–3165. [CrossRef]
16. Musavi, F.; Eberle, W.; Dunford, W.G. A Phase-Shifted Gating Technique With Simplified Current Sensing for the Semi-Bridgeless AC–DC Converter. *IEEE Trans. Veh. Technol.* **2013**, *62*, 1568–1576. [CrossRef]
17. Kim, Y.; Sung, W.; Lee, B. Comparative Performance Analysis of High Density and Efficiency PFC Topologies. *IEEE Trans. Power Electron.* **2014**, *29*, 2666–2679. [CrossRef]
18. Mohanty, P.R.; Panda, A.K. Fixed-Frequency Sliding-Mode Control Scheme Based on Current Control Manifold for Improved Dynamic Performance of Boost PFC Converter. *IEEE J. Emerg. Sel. Top. Power Electron.* **2017**, *5*, 576–586. [CrossRef]
19. Alsmadi, Y.M.; Utkin, V.; Haj-ahmed, M.A.; Xu, L. Sliding mode control of power converters: DC/DC converters. *Int. J. Control* **2018**, *91*, 2472–2493. [CrossRef]
20. Montoya, D.G.; Ramos-Paja, C.A.; Giral, R. Maximum power point tracking of photovoltaic systems based on the sliding mode control of the module admittance. *Electr. Power Syst. Res.* **2016**, *136*, 125–134. [CrossRef]
21. Sira-Ramirez, H. Nonlinear variable structure systems in sliding mode: the general case. *IEEE Trans. Autom. Control* **1989**, *34*, 1186–1188. [CrossRef]
22. Utkin, V. Variable structure systems with sliding modes. *IEEE Trans. Autom. Control* **1977**, *22*, 212–222. [CrossRef]
23. Mallik, A.; Lu, J.; Khaligh, A. Sliding Mode Control of Single-Phase Interleaved Totem-Pole PFC for Electric Vehicle Onboard Chargers. *IEEE Trans. Veh. Technol.* **2018**, *67*, 8100–8109. [CrossRef]
24. El Aroudi, A.; Martínez-Treviño, B.A.; Vidal-Idiarte, E.; Cid-Pastor, A. Fixed Switching Frequency Digital Sliding-Mode Control of DC-DC Power Supplies Loaded by Constant Power Loads with Inrush Current Limitation Capability. *Energies* **2019**, *12*, 1055. [CrossRef]
25. Rathore, N.; Fulwani, D.; Rathore, A.K.; Gautam, A.R. Adaptive Sliding Mode Based Loss-Free Resistor for Power-Factor Correction Application. *IEEE Trans. Ind. Appl.* **2019**, *55*, 4332–4343. [CrossRef]
26. Harirchi, F.; Rahmati, A.; Abrishamifar, A. Boost PFC converters with integral and double integral sliding mode control. In Proceedings of the 2011 19th Iranian Conference on Electrical Engineering, Tehran, Iran, 17–19 May 2011; pp. 1–6.
27. Sudalaimani, M.; Revathi, L.V.; Senthilkumar, N. Direct Power based Sliding Mode Control of AC-DC Converter with Reduced THD. *Tehnički Vjesnik* **2018**, *25*, 72–79.

28. Shieh, H.J.; Chen, Y.Z. A Sliding Surface-Regulated Current-Mode Pulse-Width Modulation Controller for a Digital Signal Processor-Based Single Ended Primary Inductor Converter-Type Power Factor Correction Rectifier. *Energies* **2017**, *10*, 1175. [CrossRef]
29. Chincholkar, S.H.; Jiang, W.; Chan, C. A Normalized Output Error-Based Sliding-Mode Controller for the DC–DC Cascade Boost Converter. *IEEE Trans. Circuits Syst. II Express Briefs* **2020**, *67*, 92–96. [CrossRef]
30. Sira-Ramirez, H. Sliding motions in bilinear switched networks. *IEEE Trans. Circuits Syst.* **1987**, *34*, 919–933. [CrossRef]
31. Sira-Ramirez, H.; Rios-Bolivar, M. Sliding mode control of DC-to-DC power converters via extended linearization. *IEEE Trans. Circuits Syst. I Fundam. Theory Appl.* **1994**, *41*, 652–661. [CrossRef]
32. Mejía-Ruiz, G.E.; Muñoz-Galeano, N.; López-Lezama, J.M. Modeling and development of a bridgeless PFC Boost rectifier. *Rev. Fac. Ing. Univ. Antioq.* **2017**, *82*, 9–21. [CrossRef]
33. Tan, S.; Lai, Y.M.; Tse, C.K.; Martinez-Salamero, L.; Wu, C. A Fast-Response Sliding-Mode Controller for Boost-Type Converters With a Wide Range of Operating Conditions. *IEEE Trans. Ind. Electron.* **2007**, *54*, 3276–3286. [CrossRef]
34. Siew-Chong Tan.; Lai, Y.M.; Tse, C.K. A unified approach to the design of PWM-based sliding-mode voltage controllers for basic DC-DC converters in continuous conduction mode. *IEEE Trans. Circuits Syst. I Regul. Pap.* **2006**, *53*, 1816–1827. [CrossRef]
35. Mejía-Ruiz, G.E.; Muñoz-Galeano, N.; Ortiz-Castrillón, J.R. Banda de Histéresis Adaptativa para un Convertidor AC-DC Elevador sin Puente, con Correccion del Factor de Potencia y Control por Modos Deslizantes. *Inf. Tecnol.* **2019**, *30*, 283–292. [CrossRef]

MDPI

St. Alban-Anlage 66

4052 Basel

Switzerland

Tel. +41 61 683 77 34

Fax +41 61 302 89 18

www.mdpi.com

Applied Sciences Editorial Office

E-mail: applsci@mdpi.com

www.mdpi.com/journal/applsci